SIMPLY

MATH

Editors Michael Clark, Julie Ferris, Miezan van Zyl
US Editor Karyn Gerhard
Designers Mik Gates, Jessica Tapolcai
Managing Editor Angeles Gavira
Managing Art Editor Michael Duffy
Production Editor Gillian Reid
Senior Production Controller Meskerem Berhane
Jacket Design Development Manager
Sophia M.T.T.
Jacket Designer Akiko Kato
Associate Publishing Director Liz Wheeler
Art Director Karen Self
Publishing Director Jonathan Metcalf

First American Edition, 2022
Published in the United States by DK Publishing
1745 Broadway, 20th Floor, New York, NY 10019

Copyright © 2019 Dorling Kindersley Limited
DK, a Division of Penguin Random House LLC
22 23 24 25 26 10 9 8 7 6 5 4 3 2
003–325017–Feb/2022

A catalog record for this book
is available from the Library of Congress.
ISBN 978-0-7440-4836-0

DK books are available at special discounts when
purchased in bulk for sales promotions, premiums,
fund-raising, or educational use. For details, contact:
DK Publishing Special Markets,
1745 Broadway, 20th Floor, New York, NY 10019
SpecialSales@dk.com

Printed and bound in China

For the curious
www.dk.com

This book was made with Forest
Stewardship Council™ certified
paper – one small step in DK's
commitment to a sustainable future.
For more information go to
www.dk.com/our-green-pledge

CONSULTANT
Karl Warsi taught mathematics in schools and colleges for many years. He has created bestselling textbook series for secondary-level students worldwide and is committed to inclusion in education, and the idea that people of all ages learn in different ways.

CONTRIBUTORS
Leo Ball is an Oxford physics graduate, author, and physics teacher. He also works with the Institute of Physics, Oxford University, and the Welsh Government in preparing students from under-represented backgrounds for Oxbridge entrance.

Heather Davis has taught mathematics for 30 years. She has published textbooks for Hodder Education and managed publications for the UK's Association of Teachers of Mathematics.

Julian Emsley is a math teacher and tutor on the sunny south coast of England. He enjoys growing fruit and vegetables, and biking around the hills and lanes of the South Downs.

Sue Pope is a longstanding member of the Association of Teachers of Mathematics and co-runs workshops on the history of mathematics in teaching at their conferences. Published widely, she recently coedited *Enriching Mathematics in the Primary Curriculum*.

Susan Watt studied mathematics and science at Cambridge University, and holds postgraduate degrees in philosophy and psychology. She was editor of the international magazine *Science in School* and has contributed to many math and science books for DK and other publishers.

CONTENTS

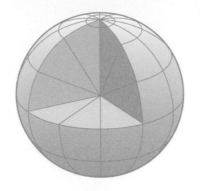

GEOMETRY

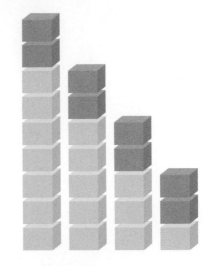

ALGEBRA

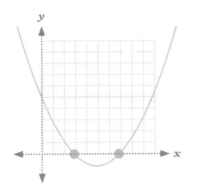

GRAPHS

RATIO AND PROPORTION

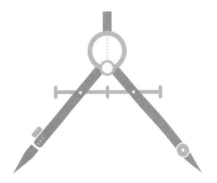

MEASURE

STATISTICS AND PROBABILITY

CALCULUS

WHAT IS MATH?

The ancient Greek philosopher Proclus once wrote,
"Wherever there is number, there is beauty." Mathematics,
or math, is regarded as beautiful by some and daunting by
others, although its remit stretches far beyond numbers into
areas such as algebra and geometry. Indeed, it is highly likely
that there are whole topics as yet undreamed of. It is difficult
to give a simple definition of math, although it is generally
accepted that its study involves quantities, shapes, and patterns.
The vast subject that we encounter today has evolved over
thousands of years, from early evidence of tallying found in
etched bones, via the invention of the Hindu-Arabic number
system, to modern abstract algebra.

   Math is vitally important, not just in passing tests, but in
many aspects of our everyday lives such as in the supermarket
or when decorating a room. Many of us have set ways of working
out a few things, though we do not always have the conceptual
understanding needed to confidently apply math in a wide
variety of situations.

   This book offers a concise and visual introduction to the big
ideas behind the main strands, such as number, measure, and
statistics. It touches on more advanced topics such as calculus,
as well as fascinating aspects including fractal geometry and
the golden ratio. It is hoped that this book will convey the beauty
of math, as well as being of some practical use, and most
importantly allow the subject to seem less daunting to many.

NUMB

E R S

A number is a symbol or word we use to represent a quantity of something. It may represent a whole number of things, a fraction, or a negative amount. Numbers may even represent so-called imaginary quantities. Representation of numbers might have started with simple tally marks made by cave dwellers but have evolved via systems such as Roman numerals to the elegant Arabic numerals in common use today. Although the Arabic system is ubiquitous, other number systems such as the Chinese system are also used in everyday mathematics.

> Integers are whole numbers and their negative opposites. Zero is also an integer.

Number line
Integers can be portrayed on a number line, stretching to infinity in both directions.

POSITIVE INTEGERS

4

3

2

1

ZERO 0

NUMBERS ABOVE

AND BELOW ZERO

-1

NEGATIVE INTEGERS

-2

-3

-4

Natural numbers (1, 2, 3…) form the foundation of mathematics. Zero is a number in its own right, and taken together the numbers 0, 1, 2, 3… to infinity are called whole numbers. Numbers less than zero (-1, -2, -3…) are called negative numbers. They are used to describe, for example, subzero temperatures. Whole numbers and their negative opposites are collectively called integers.

BREAKING NUMBERS

Numbers that lie in between integers can often be written as fractions. They are useful when measuring a quantity precisely or splitting something into equal parts. The word fraction derives from the Latin *fractio*, meaning "to break," and a fraction is a way of showing a quantity that is part of a whole number. Fractions contain two numbers, one on top of the other and separated by a bar.

NUMERATOR
The top number is the number of equal parts being described.

$$\frac{3}{4}$$

VINCULUM
Middle bar

DENOMINATOR
The bottom number is the total number of equal parts.

Part of the whole
Fractions describe numbers that exist in between the integers. One whole can be split into two halves, four quarters, and so on.

1 WHOLE									
$\frac{1}{2}$					$\frac{1}{2}$				
$\frac{1}{3}$			$\frac{1}{3}$			$\frac{1}{3}$			
$\frac{1}{4}$		$\frac{1}{4}$		$\frac{1}{4}$			$\frac{1}{4}$		
$\frac{1}{5}$		$\frac{1}{5}$		$\frac{1}{5}$		$\frac{1}{5}$		$\frac{1}{5}$	
$\frac{1}{6}$		$\frac{1}{6}$		$\frac{1}{6}$		$\frac{1}{6}$		$\frac{1}{6}$	$\frac{1}{6}$
$\frac{1}{8}$	$\frac{1}{8}$	$\frac{1}{8}$	$\frac{1}{8}$	$\frac{1}{8}$	$\frac{1}{8}$	$\frac{1}{8}$			$\frac{1}{6}$
$\frac{1}{10}$	$\frac{1}{10}$	$\frac{1}{10}$	$\frac{1}{10}$	$\frac{1}{10}$	$\frac{1}{10}$	$\frac{1}{10}$	$\frac{1}{10}$	$\frac{1}{10}$	$\frac{1}{10}$

Decimal place values

Each digit has it own place value.
Adjacent digits differ from each
other by a factor of ten.

TENS — There are 4 tens, or 40

UNITS — There are 6 units, or 6

TENTHS $\frac{1}{10}$ — 2 tenths is the same as $\frac{2}{10}$ or 0.2

HUNDREDS $\frac{1}{100}$ — 5 hundredths is the same as $\frac{5}{100}$ or 0.05

46.25

FIGURES TO THE LEFT
Whole numbers sit to the
left of the decimal point.

DECIMAL POINT

FIGURES TO THE RIGHT
Fractions sit to the right
of the decimal point.

POSITIONING THE POINT

The decimal system is the number system used throughout the world. It is based on the number 10 and, like fractions, decimals can express numbers that are not whole. Parts of the number that are more than 1 are separated from the parts that are less than 1 by a decimal point. The decimal system originated in India in the 7th century. It was developed by Islamic scholars before being adopted in Europe in the 16th century.

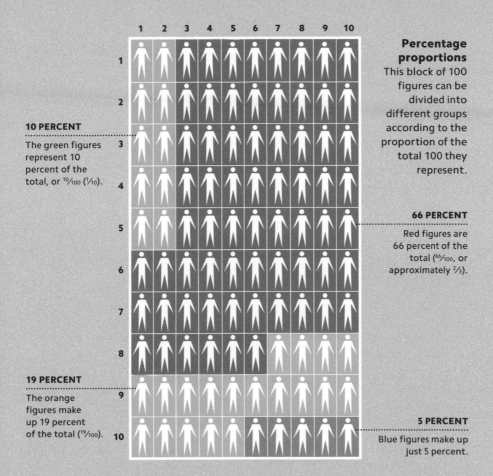

Percentage proportions
This block of 100 figures can be divided into different groups according to the proportion of the total 100 they represent.

10 PERCENT

The green figures represent 10 percent of the total, or $^{10}\!/_{100}$ ($^1\!/_{10}$).

66 PERCENT

Red figures are 66 percent of the total ($^{66}\!/_{100}$, or approximately $^2\!/_3$).

19 PERCENT

The orange figures make up 19 percent of the total ($^{19}\!/_{100}$).

5 PERCENT

Blue figures make up just 5 percent.

PARTS OF 100

We use percentages to write fractions of an amount divided into 100 parts, where 100 percent represents one whole. One percent of a quantity is one hundredth of that quantity, 50 percent is 50 hundredths, or a half, and so on. The symbol % is used to indicate a percentage. Percentages have been used in finance for more than 2000 years. Julius Caesar levied a one percent tax on the sale of goods in Ancient Rome.

THE x FACTOR

The factors of a number are the integers (whole numbers) that divide exactly into that number. For example, the factors of 12 are 1, 2, 3, 4, 6, and 12. The multiples of a number are that number multiplied by any integer. The multiples of 6 are 6, 12, 18, 24, and so on. If x is a factor of y, then y is a multiple of x, so 3 is a factor of 12 and 12 is a multiple of 3. There are a finite quantity of factors of any number, but an infinite amount of multiples.

Visualizing factors

A chocolate bar with 10 squares can be divided into equal sections in four ways. This shows the four factors of 10: 1, 2, 5, and 10.

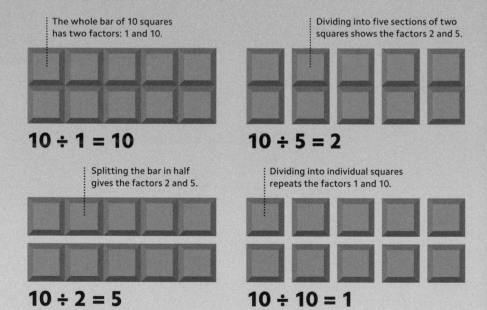

The whole bar of 10 squares has two factors: 1 and 10.

$$10 \div 1 = 10$$

Dividing into five sections of two squares shows the factors 2 and 5.

$$10 \div 5 = 2$$

Splitting the bar in half gives the factors 2 and 5.

$$10 \div 2 = 5$$

Dividing into individual squares repeats the factors 1 and 10.

$$10 \div 10 = 1$$

The only even prime is 2. All other even numbers can be divided by 2, so they are not prime.

Prime numbers are shaded green.

Up to 100
There are 25 prime numbers between 2 and 100. A number up to 100 is a prime if it cannot be divided exactly by 2, 3, 5, or 7.

1	2	3	4	5	6	7	8	9	10
11	12	13	14	15	16	17	18	19	20
21	22	23	24	25	26	27	28	29	30
31	32	33	34	35	36	37	38	39	40
41	42	43	44	45	46	47	48	49	50
51	52	53	54	55	56	57	58	59	60
61	62	63	64	65	66	67	68	69	70
71	72	73	74	75	76	77	78	79	80
81	82	83	84	85	86	87	88	89	90
91	92	93	94	95	96	97	98	99	100

PRIME TIME

A prime number is a natural number (a positive integer) that has exactly two factors, namely 1 and itself. Examples are 2, 3, 5, 7, and 13. The number 1 is not a prime since it has only one factor. Prime numbers were studied by the Ancient Greeks, including Eratosthenes, who developed an algorithm (formal method) for identifying them known as the Sieve of Eratosthenes. There are an infinite number of prime numbers and they are used in cryptography, providing security for much of the world's banking systems.

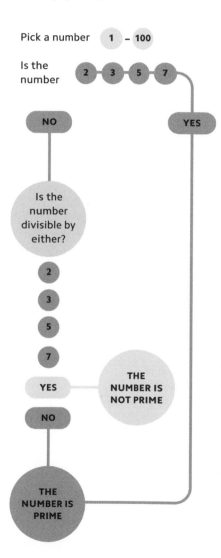

Pick a number 1 – 100

Is the number 2 3 5 7

NO — Is the number divisible by either? 2 3 5 7 — YES · THE NUMBER IS NOT PRIME — NO · THE NUMBER IS PRIME

YES — THE NUMBER IS NOT PRIME

Squares and roots

$3^2 = 9$. Conversely, the square root of a number (e.g. 9) must be multiplied by itself in order to give that number. It is written as $\sqrt{x}$. The square root of 9 is 3 because $3 \times 3 = 9$.

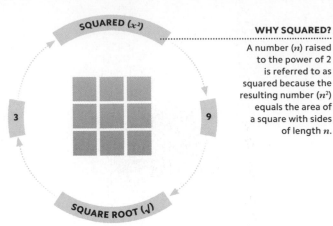

SQUARED (x^2)

SQUARE ROOT ($\sqrt{}$)

3

9

WHY SQUARED?

A number (n) raised to the power of 2 is referred to as squared because the resulting number (n^2) equals the area of a square with sides of length n.

MULTIPLIED BY ITSELF

When a number (n) is multiplied by itself x number of times, this is represented with a power or index (see p.68) written as n^x. For example, $3 \times 3 = 3^2$ (spoken as 3 raised to the power of 2). A number multiplied to the power of 2 has been squared, and to the power of 3, cubed. Conversely, the root of a number is a value that, multiplied by itself a certain number of times, gives that number. The most common types are square roots and cube roots.

Cubes and roots

$3^3 = 3 \times 3 \times 3 = 27$. Conversely, the cube root of a number (e.g. 27) appears three times in a multiplication to give that number. A cube root is written as $^3\sqrt{x}$. Here, the cube root of 27 ($^3\sqrt{27}$) is 3.

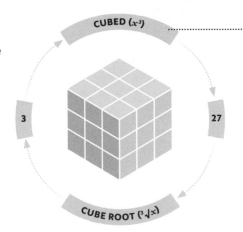

CUBED (x^3)

CUBE ROOT ($^3\sqrt{x}$)

3

27

WHY CUBED?

A number (n) raised to the power of 3 is called cubed because the resulting number (n^3) equals the volume of a cube with length, width, and height of length n.

PUZZLING POWERS

Powers are not always positive. The diagram below shows powers of 10 from 1 trillion to 1 trillionth. There is a pattern in the numbers reading down, so that as the power reduces by 1, the number is divided by 10. Because of this, $10^0=1$ and in fact any number raised to the power 0 is 1. The pattern also shows that negative powers of 10 result in numbers less than 1. When using very large or very small numbers, certain prefixes are often used to save time. A gigabyte, for example, contains 1,000,000,000 bytes of digital memory.

NAME	COMMON NOTATION	NOTATION	PREFIX
TRILLION	1 000 000 000 000	10^{12}	TERA (T)
BILLION	1 000 000 000	10^{9}	GIGA (G)
MILLION	1 000 000	10^{6}	MEGA (M)
THOUSAND	1 000	10^{3}	KILO (k)
HUNDRED	100	10^{2}	HECTO (h)
TEN	10	10^{1}	DECA (da)
ONE	1	10^{0}	[NONE]
ONE TENTH	0.1	10^{-1}	DECI (d)
ONE HUNDREDTH	0.01	10^{-2}	CENTI (c)
ONE THOUSANDTH	0.001	10^{-3}	MILLI (m)
ONE MILLIONTH	0.000 001	10^{-6}	MICRO (μ)
ONE BILLIONTH	0.000 000 001	10^{-9}	NANO (n)
ONE TRILLIONTH	0.000 000 000 001	10^{-12}	PICO (p)

Factor tree
This is a useful tool for breaking down a number into the product of its prime factors.

1. Write the number to be factorized at the top of the tree.

 56

2. Break down ("factorize") the number into two factors and write them underneath.

2   **28**

3. If a factor is a prime number then the branch ends, otherwise continue factorizing until each branch ends at a prime factor.

2   **14**

2   **7**

PRIME BUILDERS

The lowest common multiple is the lowest multiple that two or more numbers have in common. For example, the lowest common multiple of 15 and 10 is 30. The highest common factor is the largest number that two or more numbers can be divided by. For 15 and 10, the highest common factor is 5. A factor of a number that is also a prime number is called a prime factor. Any whole number greater than 1 is either a prime number or a product of prime numbers, for example $42 = 2 \times 3 \times 7$.

4. Gather the prime factors at the end of each branch.

$$2 \times 2 \times 2 \times 7 = 56$$
$$2^3 \times 7 = 56$$

5. Write the original number as a product of prime factors.

6295313 9732

Encryption
Prime factors are used to encrypt
private data such as credit card details.
A very large number, called a public
key, is made from the product of two
secret prime factors, the private key.
Only those who know the private
key can access the data.

NOTATING GREAT AND
SMALL

Standard form, which is called scientific notation in most countries, is used to write very large or very small numbers. It is based on the powers of 10 and is written as $a \times 10^b$, where a is a nonzero number between 1 and 10 and b is an integer. For example, 92,000 is $9.2 \times 10,000$, which in standard form is 9.2×10^4. This method of notation is widely used by scientists such as astronomers and cell biologists.

DNA MOLECULE WIDTH

Really small
Very small numbers have negative powers of 10 in standard form. For example, 10^{-3} is 0.001.

MICROSCOPIC

The width of a DNA molecule is about 0.000000002 meters, or 2×10^{-9} m.

Really big

To express a big integer in standard form, simply write down the first digit add a decimal point, count the number of digits after the first digit, and use that number as the power of 10.

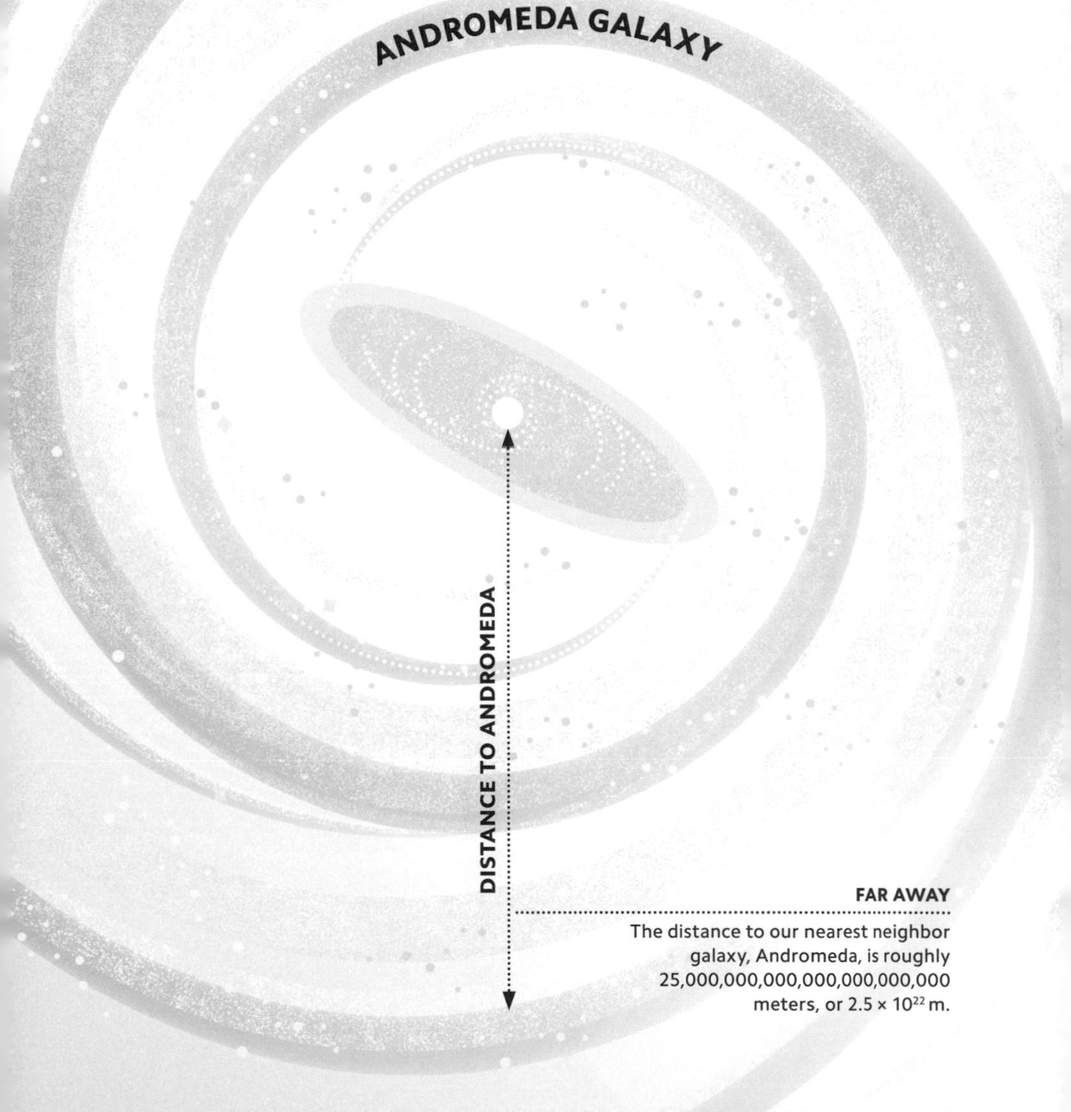

ANDROMEDA GALAXY

DISTANCE TO ANDROMEDA

FAR AWAY

The distance to our nearest neighbor galaxy, Andromeda, is roughly 25,000,000,000,000,000,000,000 meters, or 2.5×10^{22} m.

$$i^2 = \sqrt{-1}$$

IMAGINARY NUMBERS

Imagined number
No ordinary number can represent the square root of minus one, so this concept is defined with a letter, i.

A DIFFERENT WAY OF THINKING

Square any real number and the result is always positive or zero. Mathematicians invented imaginary numbers to solve problems that involve the square root of a negative number. Imaginary numbers are written using the letter i, which represents the square root of minus one. When a real number is combined with an imaginary number—for example, $4 + 3i$—it is called a complex number. Imaginary numbers are now considered an essential tool in science and engineering, for instance in the design and operation of air traffic control systems.

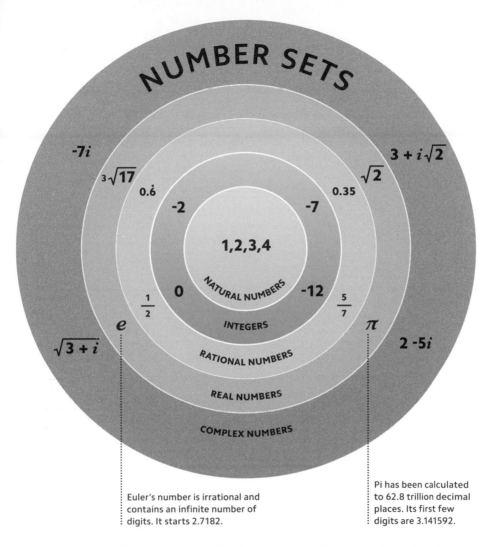

NUMBER SETS

-7*i*

$3\sqrt{17}$

0.6̇

-2

$3 + i\sqrt{2}$

$\sqrt{2}$

0.35

-7

1,2,3,4

NATURAL NUMBERS

$\frac{1}{2}$

0

-12

$\frac{5}{7}$

INTEGERS

e

π

RATIONAL NUMBERS

$\sqrt{3+i}$

2 -5*i*

REAL NUMBERS

COMPLEX NUMBERS

Euler's number is irrational and contains an infinite number of digits. It starts 2.7182.

Pi has been calculated to 62.8 trillion decimal places. Its first few digits are 3.141592.

Some number types are subsets of others, for example natural numbers are a subset of integers. On a higher level, numbers are classified as either rational or irrational. Rational numbers can be expressed as a fraction comprising two integers and irrational numbers cannot. Taken together, rational and irrational numbers form the set of real numbers. In turn, real numbers are a subset of complex numbers, which contain both real and imaginary numbers.

ON
1
8s

OFF
0
4s

ON
1
2s

OFF
0
1s

Binary number
These bulbs represent ten in binary. Adjacent digits differ by a factor of two. Add values containing 1 (place values 8 and 2) to get ten.

DIFFERENT BASES

In everyday mathematics we use base 10, where adjacent digits in a number differ from each other by a factor of 10. For example, 42 consists of the digits 4 and 2, representing 40 and 2 respectively. Our use of base 10 stems from 10 fingers on our hands. Other bases exist, particularly in computing, where base 2, also called binary notation, is used, as it closely represents the actual binary calculations performed by the computer.

GROUP RULES

A group is a set of elements, such as integers or shapes, and an operation that acts on two elements to produce a third element. For a group of integers, this operation could be addition, for example, and for a group of shapes it could be rotation. Groups have four axioms, or rules—closure, associativity, identity, and inverse (see below). Group theory is used to analyze systems that exhibit symmetry, such as the structure of molecules in chemistry.

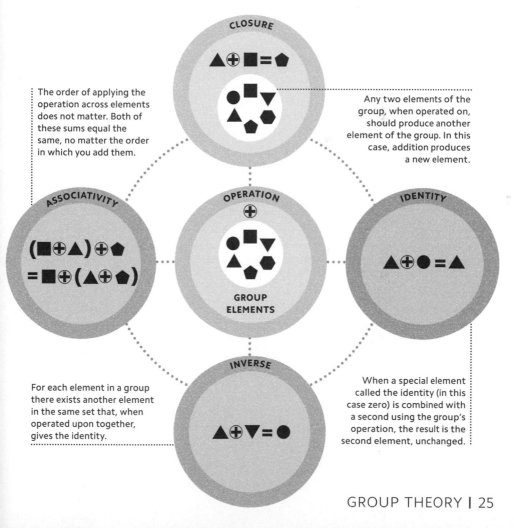

CLOSURE

The order of applying the operation across elements does not matter. Both of these sums equal the same, no matter the order in which you add them.

Any two elements of the group, when operated on, should produce another element of the group. In this case, addition produces a new element.

ASSOCIATIVITY

OPERATION

GROUP ELEMENTS

IDENTITY

For each element in a group there exists another element in the same set that, when operated upon together, gives the identity.

INVERSE

When a special element called the identity (in this case zero) is combined with a second using the group's operation, the result is the second element, unchanged.

CALCU

ATIONS

Calculation involves taking two numbers and using an operation to create an output. The main four operations are addition, subtraction, multiplication, and division. Calculations are what we do every day with the numbers we use, and understanding how to perform them is a life skill. Having strategies to work out answers in your head, or on paper, develops a quicker and accurate way to manipulate numbers in many areas of math. Calculations get more complicated with fractions or decimal values. Calculators can be used to make performing these tasks simple.

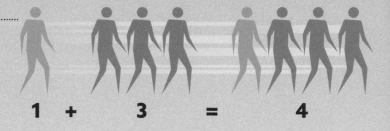

1   +   3   =   4

TWO STEPS FORWARD, ONE STEP BACK

Adding or subtracting two whole numbers is a simple calculation. They achieve opposite effects by increasing or decreasing the original amount. The order of the numbers is important for subtraction: $3 + 5 = 5 + 3$, but $5 - 3$ does not give the same result as $3 - 5$. As numbers get larger, mental strategies make the task easier. Compensation temporarily adds an amount to make a number easier to work with. Partition means dealing with the units (tens, hundreds, and so on) separately to get the answer.

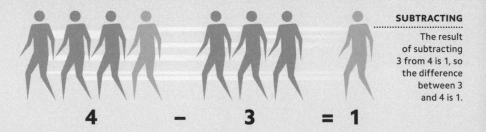

SUBTRACTING
··
The result of subtracting 3 from 4 is 1, so the difference between 3 and 4 is 1.

4   –   3   = 1

Compensation

Add an amount to one of the numbers to create a multiple of 10. This makes the sum simpler, but remember to remove the amount added afterward.

Partition

Deal with the tens and units separately to make each part of the sum easier to calculate. Then add the results of the separate sums together to solve the calculation.

35 + 18

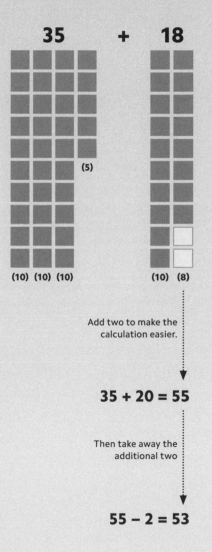

Add two to make the calculation easier.

$$35 + 20 = 55$$

Then take away the additional two

$$55 - 2 = 53$$

38 – 15

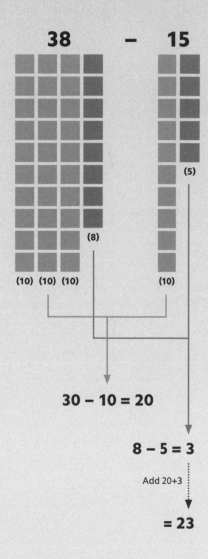

$$30 - 10 = 20$$

$$8 - 5 = 3$$

Add 20+3

$$= 23$$

How to use a multiplication table

For 6 × 7, find 6 in the first column and 7 in the top row. Then move down from the top and move across from the left side until you meet at the answer: 42.

✕	1	2	3	4	5	6	7	8	9	10
1	1	2	3	4	5	6	7	8	9	10
2	2	4	6	8	10	12	14	16	18	20
3	3	6	9	12	15	18	21	24	27	30
4	4	8	12	16	20	24	28	32	36	40
5	5	10	15	20	25	30	35	40	45	50
6	6	12	18	24	30	36	42	48	54	60
7	7	14	21	28	35	42	49	56	63	70
8	8	16	24	32	40	48	56	64	72	80
9	9	18	27	36	45	54	63	72	81	90
10	10	20	30	40	50	60	70	80	90	100

THREE THREES ARE?

Multiplication is the process of adding a number to itself a certain number of times. For example, 2 multiplied by 3 (which is written as 2 × 3) is the same as 2 + 2 + 2, and equals 6. The multiplication table above, also known as a "times table," shows the result of multiplying numbers up to 10 × 10. These can be memorized, but for large numbers a written method such as long multiplication can be used, or a calculator.

EQUAL SPLIT

Division is the opposite of multiplication. When numbers are divided, they are split into equal groups that add up to the original number. In situations where an exact split is not possible, a remainder is used. For example, if three people want to divide up ten hats equally, then the calculation required is 10 ÷ 3. The exact answer to this sum is $3\frac{1}{3}$. However, because it is impractical to split one hat into three parts, the answer is given as 3 remainder 1.

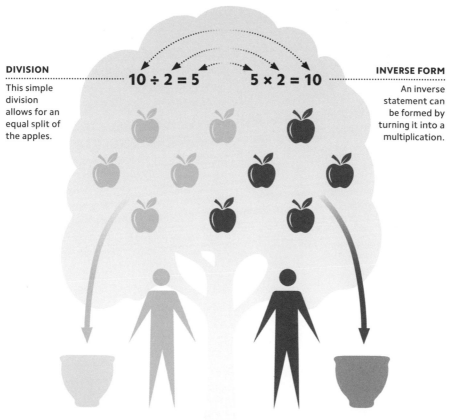

DIVISION

This simple division allows for an equal split of the apples.

10 ÷ 2 = 5   5 × 2 = 10

INVERSE FORM

An inverse statement can be formed by turning it into a multiplication.

10 APPLES ÷ 2 PEOPLE = 5 APPLES PER PERSON

HUNDREDTHS AND THOUSANDTHS

Decimal numbers use a point to indicate part of a whole number. For example, when using cash you know that $2.37 is $2 and 37 cents. The first number after the point represents tenths, the next hundredths, and so on. Decimal values can be added, subtracted, multiplied, and divided. If a decimal is multiplied by 10, the digits move one place to the left relative to the point, and they move one place to the right if it is divided by 10.

Place values

The position of a digit in a number shows the value it represents. The value of the digits increases by a factor of 10 as you go to the left, and decreases by a factor of 10 as you go to the right.

194.645 × 10

1000S THOUSANDS	100S HUNDREDS	10S TENS
	1	9
1	9	4

MULTIPLYING

MOVE TO THE LEFT

Multiplying a number by 10 means each digit represents 10 times what it did before. The effect is to move each digit one place value larger.

Multiplying decimals

To multiply decimals, simply remove the decimals, multiply the numbers, and then place the decimal point in the answer.

To convert back into decimal, simply add the number of decimal places from the first and second numbers to give the number of decimal places in the result.

12.95 × 2.6

LOSE THE DECIMAL POINTS

```
  1 2 9 5
×     2 6
─────────
3 3 6 7 0
```

```
  1 2 · 9 5 ········ 2 DECIMAL PLACES
×     2 · 6  ········ +
─────────── 1 DECIMAL PLACE
3 3 · 6 7 0 ········ 3 DECIMAL PLACES
```

1 ONES		1/10 TENTHS	1/100 HUNDREDTHS	1/1000 THOUSANDTHS
4	•	6	4	5
6	•	4	5	

DIVIDING →

Dividing a number by 10 means each digit represents a tenth of what it did before. The effect is to move it one place value smaller.

MOVE TO THE RIGHT

MORE OR LESS

Rounding is used when an exact number or measurement isn't needed and an estimate will do, or when doing a rough check of a calculation. Numbers can be rounded up or down to the nearest ten, hundred, thousand, or ten thousand and so on. Numbers that are decimals can be rounded up or down to a chosen number of decimal places. Whether a number is rounded up or down depends on where it falls within an interval of numbers. Numbers above the midpoint are rounded up, and numbers below are rounded down.

ROUNDING DOWN

Digits 1 to 4 are rounded down.

To the nearest ten

If a number is on or above the midpoint, round up. If it is below, round down. For example, 35 rounds up to 40 and 34 rounds down to 30.

31 ≈ 30

Approximately equal

An approximately equal sign is used to show a number has been rounded up or down. For example, 31 ≈ 30 and 187 ≈ 200.

ROUNDING UP

Digits 5 to 9 are rounded up.

Adding fractions
When adding, each fraction must have the same denominator. In this example they are based on quarter slices, so the numerators can simply be added.

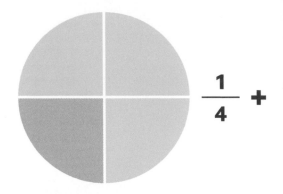

$$\frac{1}{4} +$$

FRACTION SUMS

To add or subtract fractions, it helps if the denominators (bottom numbers) are the same. If they differ, find the lowest common multiple and use it to create a common denominator (see below). Then add or subtract the numerators and the answer goes over the common denominator. Multiplying fractions is straightforward: simply multiply the numerators and the denominators and put one over the other. For division, flip the numerator and denominator on the second fraction and use the process for multiplying.

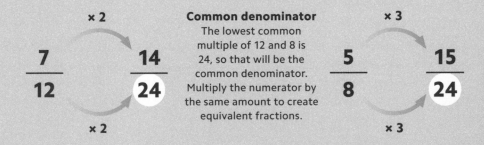

× 2

$$\frac{7}{12} \qquad \frac{14}{24}$$

× 2

Common denominator
The lowest common multiple of 12 and 8 is 24, so that will be the common denominator. Multiply the numerator by the same amount to create equivalent fractions.

× 3

$$\frac{5}{8} \qquad \frac{15}{24}$$

× 3

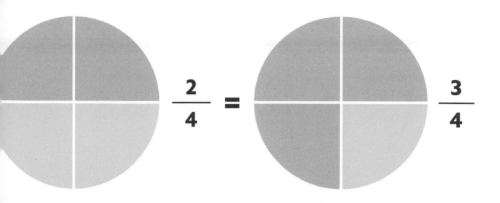

$$\frac{2}{4} = \frac{3}{4}$$

Multiple fractions

Working out the value of multiple equal fractions is the same as adding that number of fractions together.

$$\frac{1}{6} \times 2 = \qquad + \qquad = \qquad \longrightarrow$$

$$\frac{1}{6} \qquad \frac{1}{6} \qquad \frac{2}{6} \qquad \frac{1}{3}$$

Simplify the answer if the denominator and numerator have common factors.

Multiplying fractions

To multiply fractions, simply multiply the top numbers and bottom numbers separately.

1. Multiply the numerators together.

2. Multiply the denomionators.

3. Simplify the result.

$$\frac{2}{3} \times \frac{3}{4} = \frac{6}{12} = \frac{1}{2}$$

GEOM

E T R Y

Geometry is the branch of mathematics concerned with the properties and relations of points, lines, surfaces, and solids in space. The Greek mathematician Euclid's *Elements of Geometry*, based on a set of axioms, has dominated thinking about geometry up to the present day, and indeed that book was often used as a standard textbook until the early 20th century. However, from the 19th century onward, other fascinating geometry systems began to develop, partly in response to emerging disciplines such as electromagnetism and cosmology. Today, geometry is a vast field of study with wide-ranging application in everything from climate change to microbiology.

A right angle is one quarter of a whole turn, or 90°. Its two sides are perpendicular (L-shaped).

135°

90°

45°

Naming angles
Angles are named according to magnitude: acute angles are 0°–90°; obtuse angles are 90°–180°; and reflex angles are 180°–360°.

235°

MEASURING TURN

An angle happens at a point (vertex) where two straight lines meet. It is a measure of turn from one straight line to the other straight line. Angles are usually measured in degrees and one full turn is 360°. Sometimes, angles are given in a unit called a radian—a full turn is 2π (approximately 6.28) radians. Bearings are angles used to measure direction, starting at north and moving clockwise. For example, east is 090°. Bearings are always expressed in three digits.

PARALLEL RULES

Parallel lines are straight lines that are always the same distance apart. They never meet each other, no matter how far the lines are extended. When a line (called a transversal) is drawn across parallel lines, the same angles are created in several places, as the same set of vertically opposite angles (for example, a pair of blue angles or a pair of yellow angles in the diagram below) is repeated.

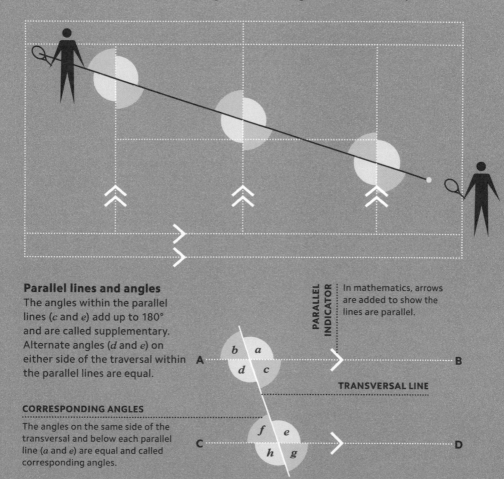

Parallel lines and angles

The angles within the parallel lines (c and e) add up to 180° and are called supplementary. Alternate angles (d and e) on either side of the traversal within the parallel lines are equal.

CORRESPONDING ANGLES

The angles on the same side of the transversal and below each parallel line (a and e) are equal and called corresponding angles.

PARALLEL INDICATOR
In mathematics, arrows are added to show the lines are parallel.

A ·········· B

b a
d c

TRANSVERSAL LINE

C ·········· D

f e
h g

FLAT SHAPES

The simplest two-dimensional shape is the triangle, which has three edges, or sides. Its stable structure means that it is used commonly in architecture. Other 2D shapes include circles and quadrilaterals (shapes with four edges). Collectively, 2D shapes with straight edges are called polygons (meaning "many sides"). Circles are 2D shapes with one curved edge (see p.105).

TRIANGLE
(3 SIDES AND ANGLES)

QUADRILATERAL
(4 SIDES AND ANGLES)

PENTAGON
(5 SIDES AND ANGLES)

HEXAGON
(6 SIDES AND ANGLES)

HEPTAGON
(7 SIDES AND ANGLES)

OCTAGON
(8 SIDES AND ANGLES)

Regular polygons

In a regular polygon, all edges are the same length and the angles are always equal in size. As the number of edges increases in a regular polygon, the shape gets closer to resembling a circle.

NONAGON
(9 SIDES AND ANGLES)

DECAGON
(10 SIDES AND ANGLES)

HENDECAGON
(11 SIDES AND ANGLES)

DODECAGON
(12 SIDES AND ANGLES)

PENTADECAGON
(15 SIDES AND ANGLES)

ICOSAGON
(20 SIDES AND ANGLES)

FLAT WITH THREE SIDES

The triangle is the simplest polygon and it is the only polygon that is rigid. Triangles therefore abound in construction, for example electricity pylons and roof trusses. Triangles are defined by the lengths of their sides and the angles at the vertices (corners).

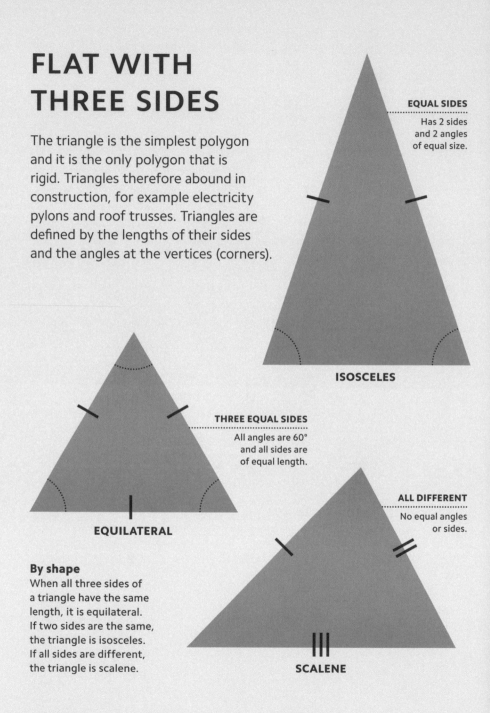

EQUAL SIDES
Has 2 sides and 2 angles of equal size.

ISOSCELES

THREE EQUAL SIDES
All angles are 60° and all sides are of equal length.

EQUILATERAL

ALL DIFFERENT
No equal angles or sides.

SCALENE

By shape
When all three sides of a triangle have the same length, it is equilateral. If two sides are the same, the triangle is isosceles. If all sides are different, the triangle is scalene.

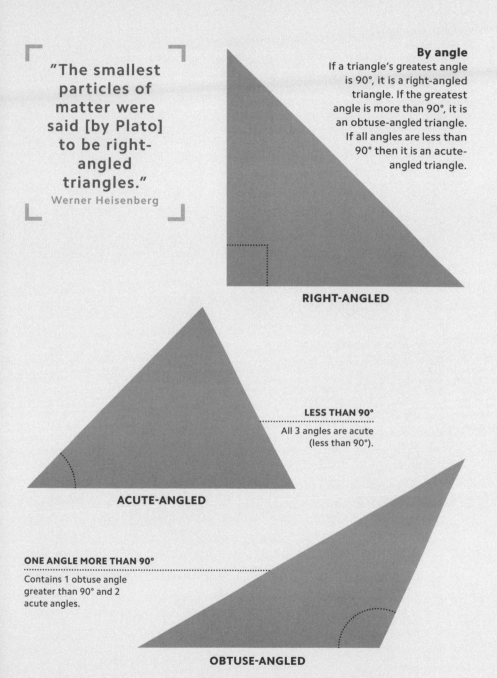

"The smallest particles of matter were said [by Plato] to be right-angled triangles."
Werner Heisenberg

By angle
If a triangle's greatest angle is 90°, it is a right-angled triangle. If the greatest angle is more than 90°, it is an obtuse-angled triangle. If all angles are less than 90° then it is an acute-angled triangle.

RIGHT-ANGLED

ACUTE-ANGLED

LESS THAN 90°
All 3 angles are acute (less than 90°).

ONE ANGLE MORE THAN 90°
Contains 1 obtuse angle greater than 90° and 2 acute angles.

OBTUSE-ANGLED

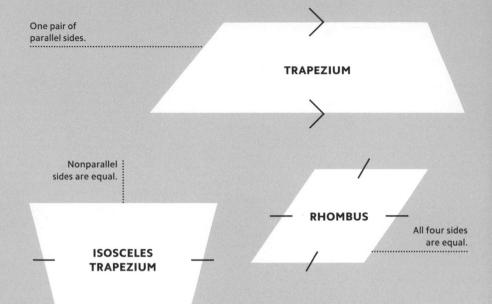

One pair of
parallel sides.

TRAPEZIUM

Nonparallel
sides are equal.

RHOMBUS

All four sides
are equal.

**ISOSCELES
TRAPEZIUM**

FOUR-SIDED FIGURES

Quadrilaterals are flat shapes with four
straight edges and four vertices (corners). Quadrilaterals
are classified by the properties of their edges and angles.
For example, a rhombus has equal sides and a square is a
special rhombus in that all angles are equal. A square is also
a special rectangle in that all edges have the same length.
The word oblong is sometimes used for a rectangle that
is not a square. Any quadrilateral can be split into
at least two triangles, so the angle sum of
a quadrilateral is always 360°.

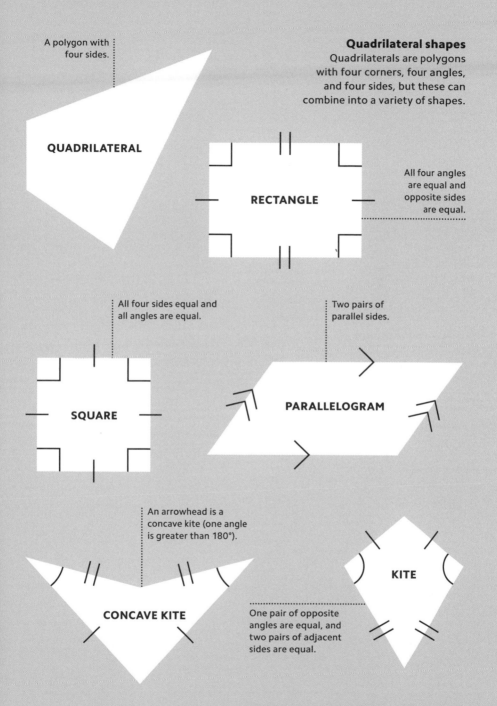

A polygon with four sides.

Quadrilateral shapes
Quadrilaterals are polygons with four corners, four angles, and four sides, but these can combine into a variety of shapes.

QUADRILATERAL

RECTANGLE

All four angles are equal and opposite sides are equal.

All four sides equal and all angles are equal.

Two pairs of parallel sides.

SQUARE

PARALLELOGRAM

An arrowhead is a concave kite (one angle is greater than 180°).

KITE

CONCAVE KITE

One pair of opposite angles are equal, and two pairs of adjacent sides are equal.

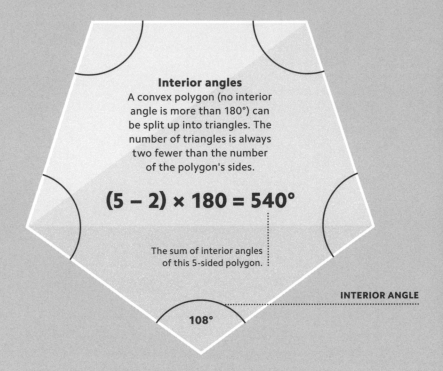

Interior angles
A convex polygon (no interior angle is more than 180°) can be split up into triangles. The number of triangles is always two fewer than the number of the polygon's sides.

$$(5 - 2) \times 180 = 540°$$

The sum of interior angles of this 5-sided polygon.

INTERIOR ANGLE

108°

THE SUM OF MANY ANGLES

Any polygon can be divided into triangles, and finding the least number of triangles allows you to determine the sum of the interior angles in the polygon. As angles in a triangle add up to 180°, the sum of the interior angles will be the least number of triangles multiplied by 180°. For a quadrilateral (see pp.46–47), the sum of the interior angles is 360°, for a pentagon it is 540°, and so on. An exterior angle is formed on the outside of a polygon when one of its sides is extended. If you imagine walking around any polygon, you will turn a total of 360°, or one full turn, to get back to where you started. The sum of the exterior angles of a polygon is therefore always 360°.

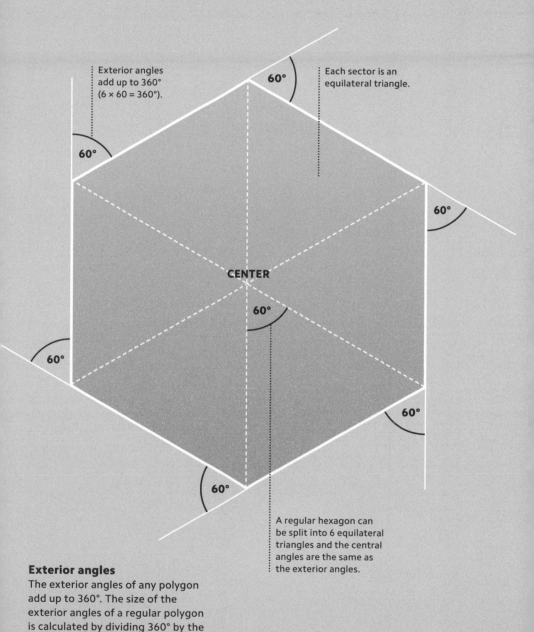

Exterior angles
add up to 360°
(6 × 60 = 360°).

Each sector is an
equilateral triangle.

60°

60°

60°

CENTER

60°

60°

60°

60°

A regular hexagon can
be split into 6 equilateral
triangles and the central
angles are the same as
the exterior angles.

Exterior angles

The exterior angles of any polygon
add up to 360°. The size of the
exterior angles of a regular polygon
is calculated by dividing 360° by the
number of sides the polygon has.

3D SHAPES

Shapes that occupy space are three-dimensional (3D). Polyhedra are 3D shapes that have flat faces. The simplest polyhedron has four triangular faces. Not all 3D shapes are polyhedra; the sphere comprises all the points equidistant from a single point, creating a single curved surface. Cones and cylinders have both curved and flat faces. The net of a 3D shape occurs when the shape is "laid flat"—easy for polyhedra but impossible to do exactly for a sphere, which is why world atlases are always slightly distorted.

TETRAHEDRON
This shape is composed of equilateral triangles. The regular tetrahedron has 4 faces, 6 edges, and 4 vertices.

CUBE
A six-sided prism made up of rectangular faces is a cuboid. A cube is a type of cuboid with all edges of equal lengths, and six square faces.

OCTAHEDRON
A regular octahedron is composed of equilateral triangles. It has 8 faces, 12 edges, and 6 vertices.

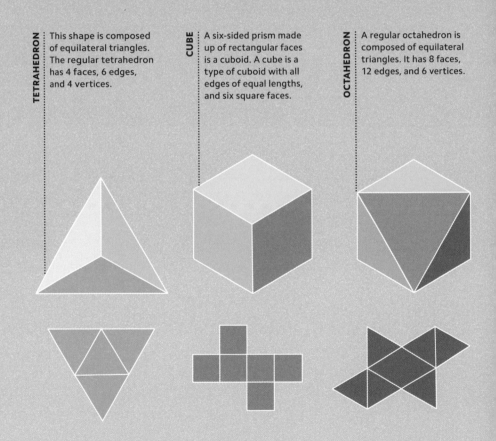

> Each Platonic solid was linked with an element:
> tetrahedron—fire
> cube—earth
> octahedron—air
> dodecahedron—
> the universe
> icosahedron—water.

Platonic solids

There are just five regular polyhedra, known as Platonic solids after the Greek philosopher Plato. A regular polyhedron has identical regular polygon faces with the same number of faces meeting at each vertex (point where edges meet).

DODECAHEDRON

This solid is composed of regular pentagons. The regular dodecahedron has 12 faces, 30 edges, and 20 vertices.

ICOSAHEDRON

The regular icosahedron has 20 faces, 30 edges, and 12 vertices. It is composed of equilateral triangles (see pp.44–45).

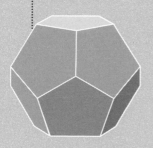

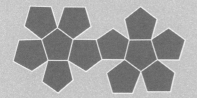

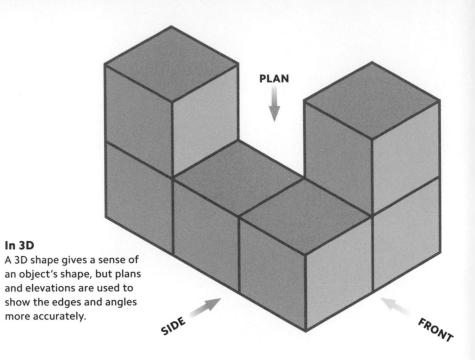

PLAN

SIDE

FRONT

In 3D

A 3D shape gives a sense of an object's shape, but plans and elevations are used to show the edges and angles more accurately.

MAKING PLANS

Plans and elevations are ways of accurately representing 3D objects in 2D. The plan shows the object from directly above and the elevations show what each side of the object looks like when viewed directly. Another way to show 3D shapes in 2D is to use the net (see p.50), which shows all the faces of the 3D shape and how they are folded up to make the solid.

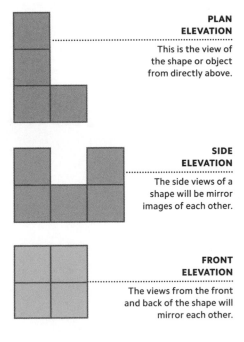

PLAN ELEVATION

This is the view of the shape or object from directly above.

SIDE ELEVATION

The side views of a shape will be mirror images of each other.

FRONT ELEVATION

The views from the front and back of the shape will mirror each other.

Lines of symmetry

A 2D shape has reflective symmetry when it can be divided by a line into two identical parts.

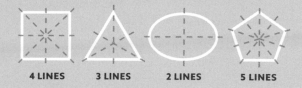

4 LINES **3 LINES** **2 LINES** **5 LINES**

MULTIPLE COPIES

Symmetry is a property that means something looks the same even when it is transformed. Rotational symmetry occurs when a shape or object can be rotated around its center and fits into the original outline of the shape more than just once. The number of times the shape fits into its original outline is called the order. For example, a square has order of rotational symmetry four. Reflective symmetry occurs when an object or shape can be divided into identical parts.

Cuboid symmetry

As well as rotational symmetry, 3D objects may have planes of reflective symmetry rather than lines.

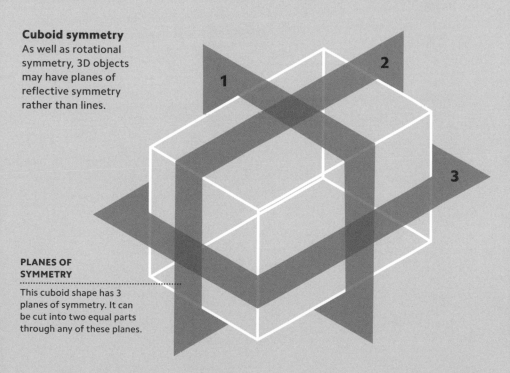

PLANES OF SYMMETRY

This cuboid shape has 3 planes of symmetry. It can be cut into two equal parts through any of these planes.

Geographical coordinates

Coordinates on maps use a grid of latitude and longitude. Latitude gives the position north or south of the equator, and longitude gives the position east or west of the prime meridian.

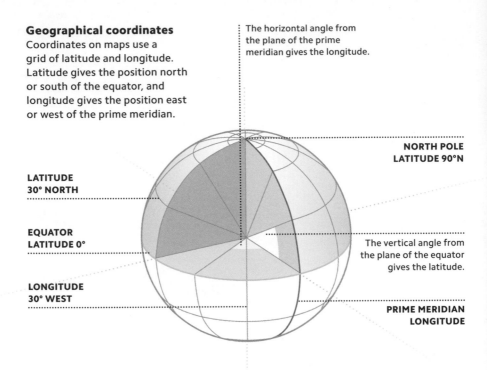

The horizontal angle from the plane of the prime meridian gives the longitude.

NORTH POLE LATITUDE 90°N

LATITUDE 30° NORTH

EQUATOR LATITUDE 0°

The vertical angle from the plane of the equator gives the latitude.

LONGITUDE 30° WEST

PRIME MERIDIAN LONGITUDE

Coordinate geometry is also known as Cartesian geometry, after its inventor, the French philosopher René Descartes. It uses two coordinate axes: two lines drawn at right angles. The intersection of the axes is called the origin. Every point has a unique coordinate that says how far it is from each axis. Coordinate geometry allows points, lines, and curves to be described by algebra (see pp.64–77) and manipulated using algebraic techniques. It also allows us to visualize algebra. For example, the equation $x^2 + y^2 = 1$ is true for every point on the grid that is one unit away from the origin and is plotted as a circle.

PLOTTING POINTS

3D Cartesian grid

A pair of 2D coordinates specifies a point by its distance from the origin (the point where the axes intersect) along the x-axis (horizontal) and y-axis (vertical). The coordinates are written as a pair of numbers in brackets and separated by a comma. Adding another (z) axis means a third dimension can also be plotted.

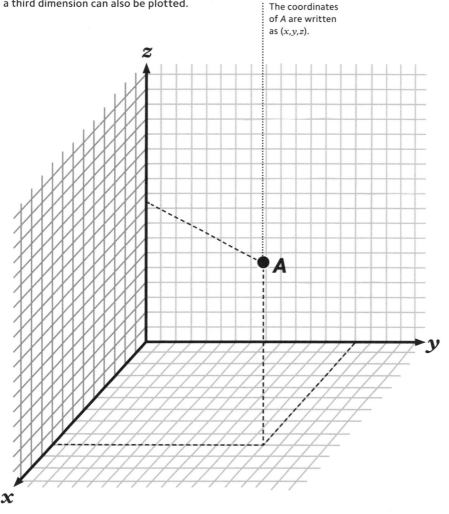

The coordinates of A are written as (x,y,z).

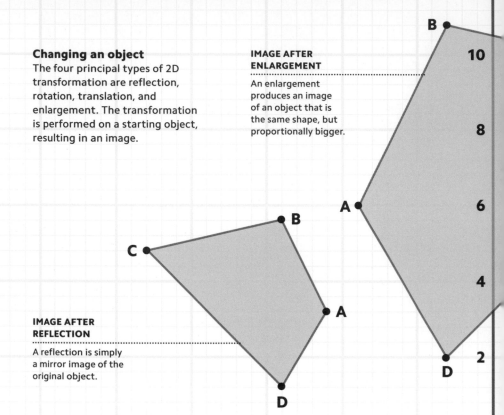

Changing an object
The four principal types of 2D transformation are reflection, rotation, translation, and enlargement. The transformation is performed on a starting object, resulting in an image.

IMAGE AFTER ENLARGEMENT
An enlargement produces an image of an object that is the same shape, but proportionally bigger.

IMAGE AFTER REFLECTION
A reflection is simply a mirror image of the original object.

-12 -10 -8 -6 -4 -2 0

10

8

6

4

2

-2

-4

-6

-8

TRANSFORMING SHAPES

Transformation means change. In geometry, it means change applied to a shape, or object, resulting in an image. Three main types of transformation are reflection, rotation, and translation. These all result in an image that is congruent to the object, meaning it is identical in shape and size. Another type of transformation, enlargement, results in an image that is similar to the object: an identical shape but different size. Transformations are commonly used in computer graphics.

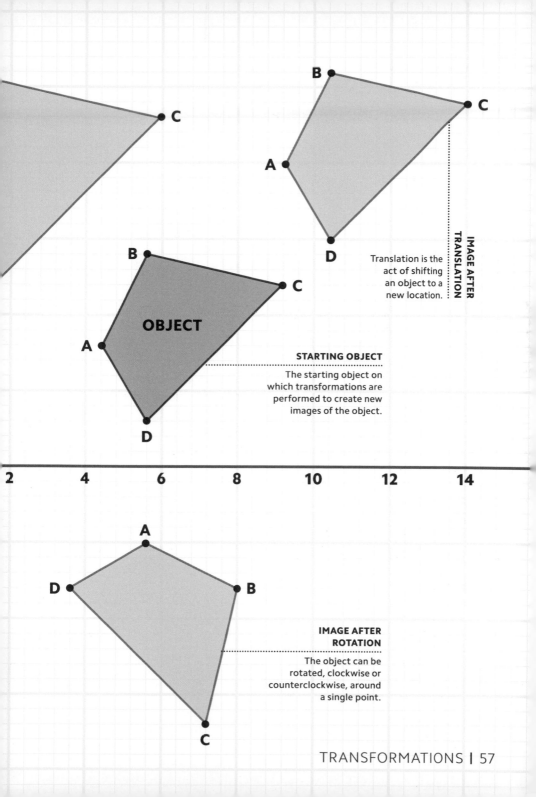

C

B

C

A

D

IMAGE AFTER TRANSLATION

Translation is the act of shifting an object to a new location.

B

C

OBJECT

A

D

STARTING OBJECT

The starting object on which transformations are performed to create new images of the object.

| 2 | 4 | 6 | 8 | 10 | 12 | 14 |

A

D

B

C

IMAGE AFTER ROTATION

The object can be rotated, clockwise or counterclockwise, around a single point.

TRANSFORMATIONS | 57

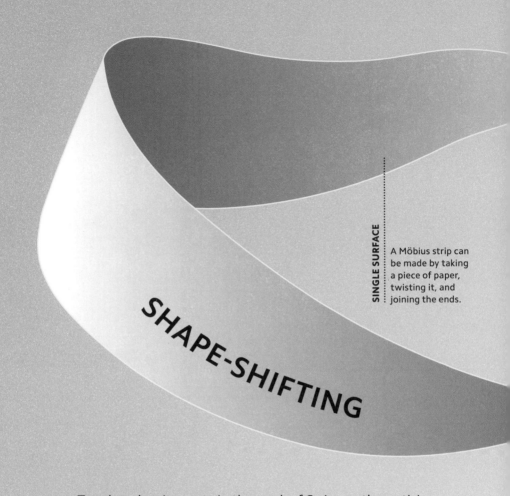

A Möbius strip can be made by taking a piece of paper, twisting it, and joining the ends.

SHAPE-SHIFTING

Topology has its roots in the work of Swiss mathematician Leonhard Euler on many-sided shapes. It is known as "rubber sheet" geometry, because the idea is that the geometric objects can be imagined to be on a rubber sheet that can be stretched and bent, but cutting and gluing is not allowed. These shapes can, therefore, be thought of as being without measurements, such as length, proportion, or angles. In topology, the relative positions of objects are important, but not the distances and angles between objects.

Similar shapes

A mug and a doughnut shape (which is called a torus) are equivalent, as one shape can be continuously deformed into the other.

NOT A HOLE

The cylinder has only one open end, so it can be filled in. In the same way, topologically speaking, a glass and a plate are equivalent.

SHAPING THE TORUS

Once the mug's cylinder is filled in, the shape can be molded. The handle of the mug is kept to form the torus.

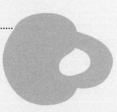

Möbius strip

This geometric shape, named after German mathematician August Möbius, has the unusual characteristics of having only one side and one continuous edge. There is no conventional "inside" or "outside" part of the shape. It can be stretched or twisted in any direction, and it will still only have a single surface.

SAME SHAPE

Neither the mug nor the donut can be deformed into a figure of 8, as this shape has two holes.

4D GEOMETRY

In the 1800s, mathematician Hermann Minkowski added time to 3D space to create 4D "spacetime," which is necessary for understanding Einstein's theory of special relativity. Four values define an event in Minkowski space—three for the location in 3D and one for the time at which the event occurred. Time is a relative quantity, running at different speeds for two people traveling at different speeds relative to each other. However, the speed of light is constant and this gives rise to constraints on the future and past events associated with any point in spacetime.

THINGS TO COME

Successive events are constrained within this cone.

" ...space by itself, and time by itself, are doomed to fade away into mere shadows, and only a kind of union of the two will preserve an independent reality."

Hermann Minkowski

Spacetime

According to Minkowski, there are three axes of space, plus the axis of time, which frame cones within which all events occur.

FUTURE LIGHT CONE

TIME

EVENT

A unique point in Minkowski spacetime, with all potential befores and afters constrained by the light cones.

SPACE

SPACE

PAST LIGHT CONE

SPACE

HYPERSURFACE

In Minkowski space, the hypersurface represents the present.

SPACE

IN THE PAST

Preceding events are constrained within this cone.

ZOOMING IN

Much of the natural world shows self-similarity (made up of smaller, similar shapes). A tree has a trunk that divides into branches, but the branches also divide into smaller branches, and then twigs. A fractal shape shows an idealized form of self-similarity: zooming in reveals smaller copies of the same fractal shape. A fractal can have a simple mathematical method for its creation, yet its shape is surprisingly complex. This means that fractals can be used to simulate and study natural objects, such as mountains, coastlines, and trees.

Endless iterations
This shape is called "Koch's snowflake," in the ultimate limit it has an infinitely long perimeter but a finite area (see p.104).

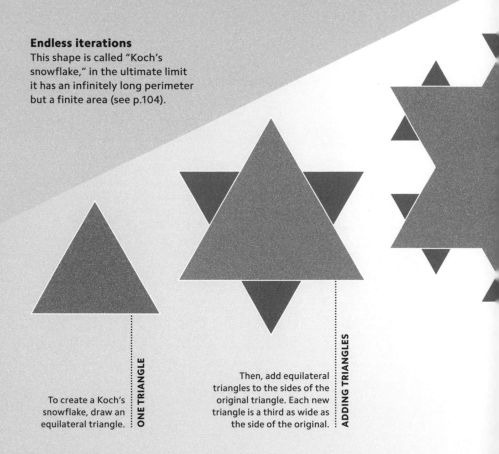

ONE TRIANGLE

To create a Koch's snowflake, draw an equilateral triangle.

ADDING TRIANGLES

Then, add equilateral triangles to the sides of the original triangle. Each new triangle is a third as wide as the side of the original.

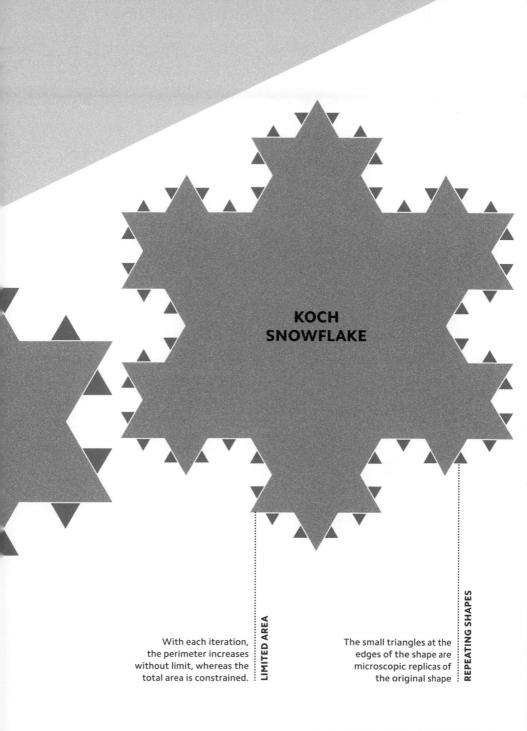

KOCH
SNOWFLAKE

With each iteration,
the perimeter increases
without limit, whereas the
total area is constrained.

LIMITED AREA

The small triangles at the
edges of the shape are
microscopic replicas of
the original shape

REPEATING SHAPES

ALGE

B R A

Algebra is derived from the Arabic *al-jabr*, which means "completion" or "reunion of broken parts." It is a simple and elegant way of representing unknowns (symbols or letters) in mathematical equations. These can either be manipulated to solve the equation, or used to define the relationship between two or more variables. Algebraic notation has many practical uses, particularly in finance, science, and engineering. Creating algebraic descriptions of the relationship between variables allows theories to be tested and adapted.

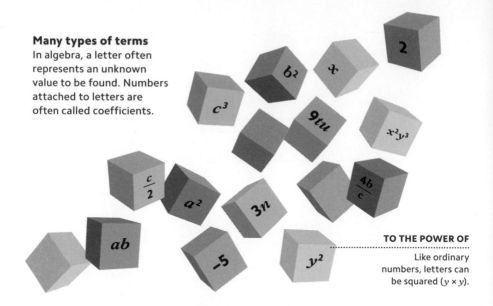

Many types of terms
In algebra, a letter often represents an unknown value to be found. Numbers attached to letters are often called coefficients.

TO THE POWER OF
Like ordinary numbers, letters can be squared ($y \times y$).

BUILDING BLOCKS

Algebra is the language of mathematics. Like any language, it has the equivalent of words, phrases, and rules. Terms act like words, expressions act like phrases, and the rules govern how they interact. Each algebraic term is either a number, a variable (a letter representing an unknown value), or a mix of the two. For example, the term $6x$ signifies 6 amounts of the variable x. When terms are combined by addition or subtraction they form an expression like the one below.

A jumble of terms
This expression is made up of the terms $6x$, 7, x, $2ab$, 2, y, and $5ab$.

In this kind of term, each part is multiplied together. In other words, $5ab = 5 \times a \times b$.

MULTIPLE VARIABLES

KEEP IT SIMPLE

Simplifying an expression is one of the most basic techniques in algebra and a helpful way to change a complex expression into a more compact form. This is often done by collecting (adding or subtracting) like terms, for example $6x - x$, or $7 - 5$. Algebraic fractions with common terms on the numerator (top) and denominator (bottom) can also be simplified by dividing by common numbers or variables.

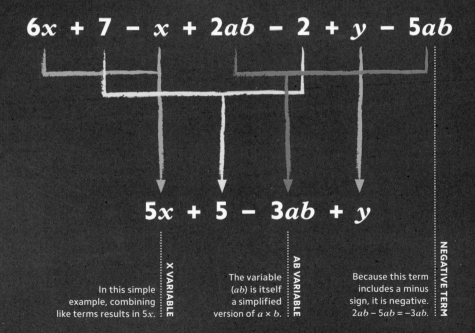

$$6x + 7 - x + 2ab - 2 + y - 5ab$$

$$5x + 5 - 3ab + y$$

X VARIABLE
In this simple example, combining like terms results in $5x$.

AB VARIABLE
The variable (ab) is itself a simplified version of $a \times b$.

NEGATIVE TERM
Because this term includes a minus sign, it is negative. $2ab - 5ab = -3ab$.

Which terms to simplify
Collecting like terms requires that they share the same form. For example, $6x - x$ gives us $5x$, but $6x + y$ cannot be simplified.

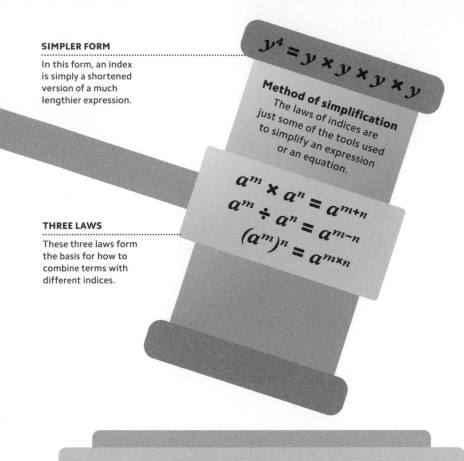

In this form, an index is simply a shortened version of a much lengthier expression.

$$y^4 = y \times y \times y \times y$$

Method of simplification

The laws of indices are just some of the tools used to simplify an expression or an equation.

$$a^m \times a^n = a^{m+n}$$
$$a^m \div a^n = a^{m-n}$$
$$(a^m)^n = a^{m \times n}$$

THREE LAWS

These three laws form the basis for how to combine terms with different indices.

USING POWERS

An index, or power, is a small floating number next to a term that indicates the number of times that term appears in a multiplication. The plural of index is indices. Indices tell us how many times to multiply a term by itself. A term with an index can be combined with like terms (other terms with the same base). There are three simple laws for combining terms with different indices (see above). To multiply, add the indices. To divide, subtract the indices. Finally, to raise a term with a power to another power, multiply the indices.

RESHAPING EXPRESSIONS

Expanding (or removing) brackets in an expression is a crucial skill in algebra. The reverse process, which creates brackets, is called factorizing. The difficulty of these processes depends on the complexity of the expression. The example below is straightforward. An expression featuring double brackets, however, is trickier. An example of this is $(x + 3)(x + 4)$. To expand double brackets, multiply every term in the first bracket by every term in the second. This gives $x^2 + 7x + 12$.

EXPANDING →

Expanding brackets removes them.

$$2(x+y) = 2x + 2y$$

Multiply everything in the brackets by 2.

2 is a common factor of the x and y terms.

← **FACTORIZING**

Transforming expressions
To expand brackets, multiply what is inside the brackets by what is outside of them. To factorize, find the greatest common factor that the terms $2x$ and $2y$ share (2), separate it, and place the rest of the expression in the brackets.

DEFINING RELATIONSHIPS

A formula (plural: formulae) describes the relationship between two or more variables. To calculate an unknown value, substitute a known value into the formula. For example, a temperature in Celsius (°C) can be converted into Fahrenheit (°F) using the formula below. Formulae feature in a variety of mathematical areas and are useful for calculating many different things, including the speed of a moving object, the dimensions of a triangle, and the volumes of different shapes (see p.106–107).

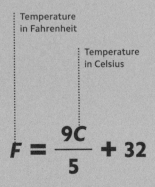

Temperature in Fahrenheit

Temperature in Celsius

$$F = \frac{9C}{5} + 32$$

Converting a temperature
Say 25°C is being coverted to °F. This formula says to multiply 25 by 9 (225), divide that by 5 (45), then add 32, resulting in 77°F.

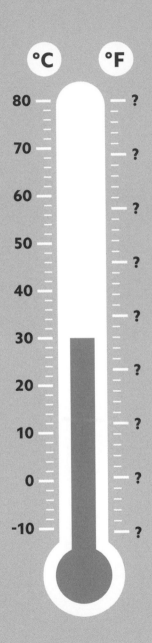

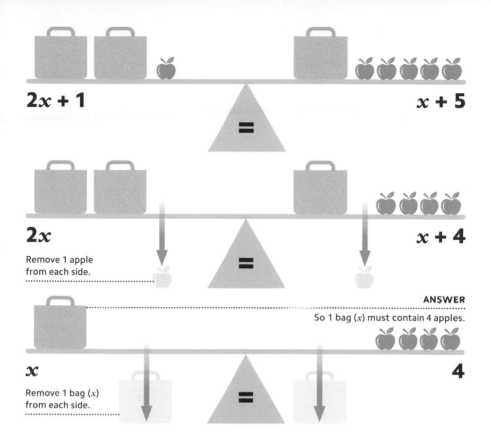

$2x + 1$

$x + 5$

$2x$

Remove 1 apple
from each side.

$x + 4$

ANSWER
So 1 bag (x) must contain 4 apples.

x

Remove 1 bag (x)
from each side.

4

Equal balance
As long as we subtract the same amount from
each side, the equation is balanced, and we can
work out how many apples are in one bag (x).

A BALANCING ACT

An equation has two expressions separated by an equals sign.
The simplest type is a linear equation, which contains one variable
along with some numbers. A simple example, such as $x + 3 = 5$, yields
an obvious value for x. With something more complex, a balance
method can be used, where we find x by adding, subtracting,
multiplying, or dividing by the same amount from each side.

Flight path
Substituting values for x into the quadratic function below gives coordinates to create this graph, which describes the flight of a ball. The ball hits the ground ($y = 0$) when x is approximately 24.8 yards.

$$y = -\tfrac{1}{4}x^2 + 6x + 5$$

HEIGHT (YARDS)

DISTANCE (YARDS)

WHAT ARE QUADRATIC EQUATIONS?

The trajectory of a projectile such as a ball or rocket can be described using a quadratic function (see pp.80–81), which can be written in the form $y = ax^2 + bx + c$. It always has a squared term as its highest power. To calculate where a projectile is likely to land, give y the value 0, so that $ax^2 + bx + c = 0$. This turns the function into an equation that can be solved. A simple way to solve quadratic equations is to plot the equation on a graph, as above.

SOLVING BY SUBSTITUTION

An equation with only one variable (such as $y + 3 = 5$) is easy to solve. One with two variables (such as such as $x + 2y = 10$) requires the use of a second equation to solve it. Two equations with the same variables are called simultaneous equations. To find x and y, the equations are solved together, by elimination or substitution. The substitution method involves rearranging one equation so that its subject (left side) consists of a single variable. This equation is substituted into the second equation.

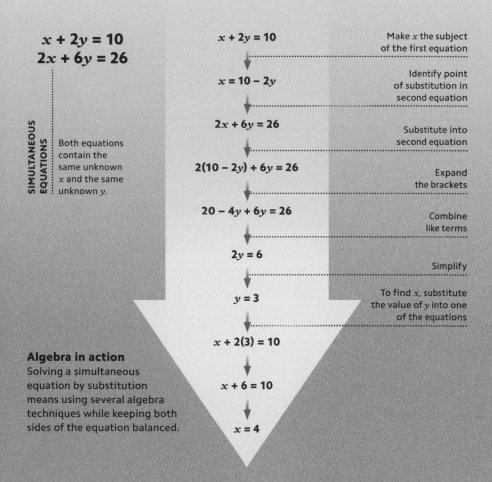

$x + 2y = 10$
$2x + 6y = 26$

SIMULTANEOUS EQUATIONS

Both equations contain the same unknown x and the same unknown y.

$x + 2y = 10$ — Make x the subject of the first equation

$x = 10 - 2y$ — Identify point of substitution in second equation

$2x + 6y = 26$ — Substitute into second equation

$2(10 - 2y) + 6y = 26$ — Expand the brackets

$20 - 4y + 6y = 26$ — Combine like terms

$2y = 6$ — Simplify

$y = 3$ — To find x, substitute the value of y into one of the equations

$x + 2(3) = 10$

$x + 6 = 10$

$x = 4$

Algebra in action
Solving a simultaneous equation by substitution means using several algebra techniques while keeping both sides of the equation balanced.

NOT ALL EQUATIONS ARE EQUAL

In an equation, an equals sign shows that one side is equal to the other side. When the two sides are not equal, the relationship is an inequality rather than an equation. An inequality can be a statement of fact: for example, it is clear that 3 is less than 5. However, in an inequality that includes a variable (such as x), there can be a range of possible values for x. There are five symbols that are used to express different types of inequality (see below).

$x \neq y$
NOT EQUAL TO

$x > y$
GREATER THAN

$x \geq y$
GREATER THAN OR EQUAL TO

$x < y$
LESS THAN

$x \leq y$
LESS THAN OR EQUAL TO

Number lines

A number line is a way of representing an inequality. It helps us to visualize the range of possible values of a variable.

$x \leq 3$

$x > 5$

The circle is filled in to show that 3 is a possible value of x.

The blue line indicates that x is any value less than or equal to 3.

The green line denotes that x is any value greater than 5.

Because x can be any value greater than 5, but not 5 itself, this circle is left unfilled.

-3 -2 -1 0 1 2 3 4 5 6 7 8

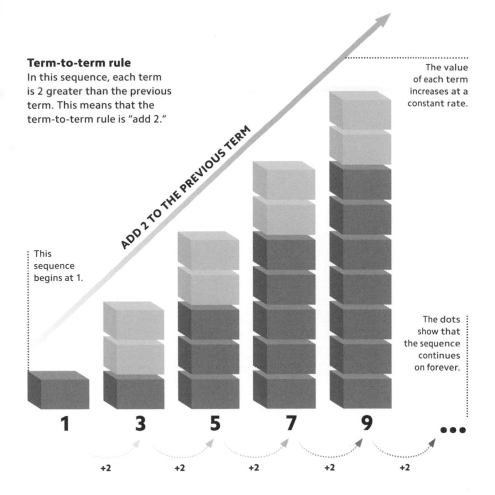

Term-to-term rule
In this sequence, each term is 2 greater than the previous term. This means that the term-to-term rule is "add 2."

ADD 2 TO THE PREVIOUS TERM

The value of each term increases at a constant rate.

This sequence begins at 1.

The dots show that the sequence continues on forever.

1   3   5   7   9   •••

+2   +2   +2   +2   +2

WHAT NEXT?

A sequence is a list of terms that follow a specific rule or rules. We can describe a sequence by identifying the term-to-term rule. This is the rule for working out the next term in the sequence. In a linear sequence, we work out the next term by adding or subtracting the same amount each time. In a geometric sequence, multiplication or division by the same amount each time is used instead.

55

Fibonacci also introduced Arabic numbers (1, 2, 3, etc.) to Europe, replacing Roman numerals.

GOING FOR GOLD

A number in the Fibonacci sequence divided by the previous number creates a ratio. As the sequence progresses, this ratio becomes closer and closer to the golden ratio (see pp.96–97).

UNIQUE NUMBER PATTERNS

While linear and geometric sequences follow basic rules (see p.75), other sequences can be slightly more complex. These include square and cube number sequences, where each term is squared or cubed, and the Fibonacci sequence (see opposite). The first two terms of the Fibonacci sequence are 1 and 1, after which each term is the sum of the two previous terms. This can be visualized as a line spiraling outward as it grows.

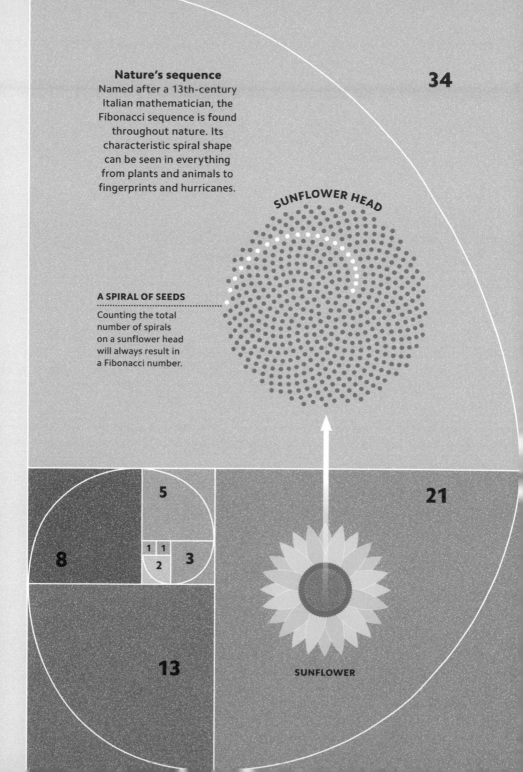

Nature's sequence
Named after a 13th-century Italian mathematician, the Fibonacci sequence is found throughout nature. Its characteristic spiral shape can be seen in everything from plants and animals to fingerprints and hurricanes.

34

SUNFLOWER HEAD

A SPIRAL OF SEEDS
Counting the total number of spirals on a sunflower head will always result in a Fibonacci number.

5

8

1 1
2

3

21

13

SUNFLOWER

GRAP

H   S

Graphs are a visual representation of the relationship
between two or more things. In two dimensions, this can
be the plot of one variable against another. A function
defines the relationship between the variables, and
the plot contains all points that satisfy it. Cartesian
coordinates (x and y) are often used to represent the
points plotted against the x- and y-axis, which intersect at
the origin (0, 0). Graphs are used to help solve real-world
problems, such as displaying information gathered from
experiments or tracking changes over time. More complex
3D graphs can be created if three variables are used.

FUNCTION

MACHINES

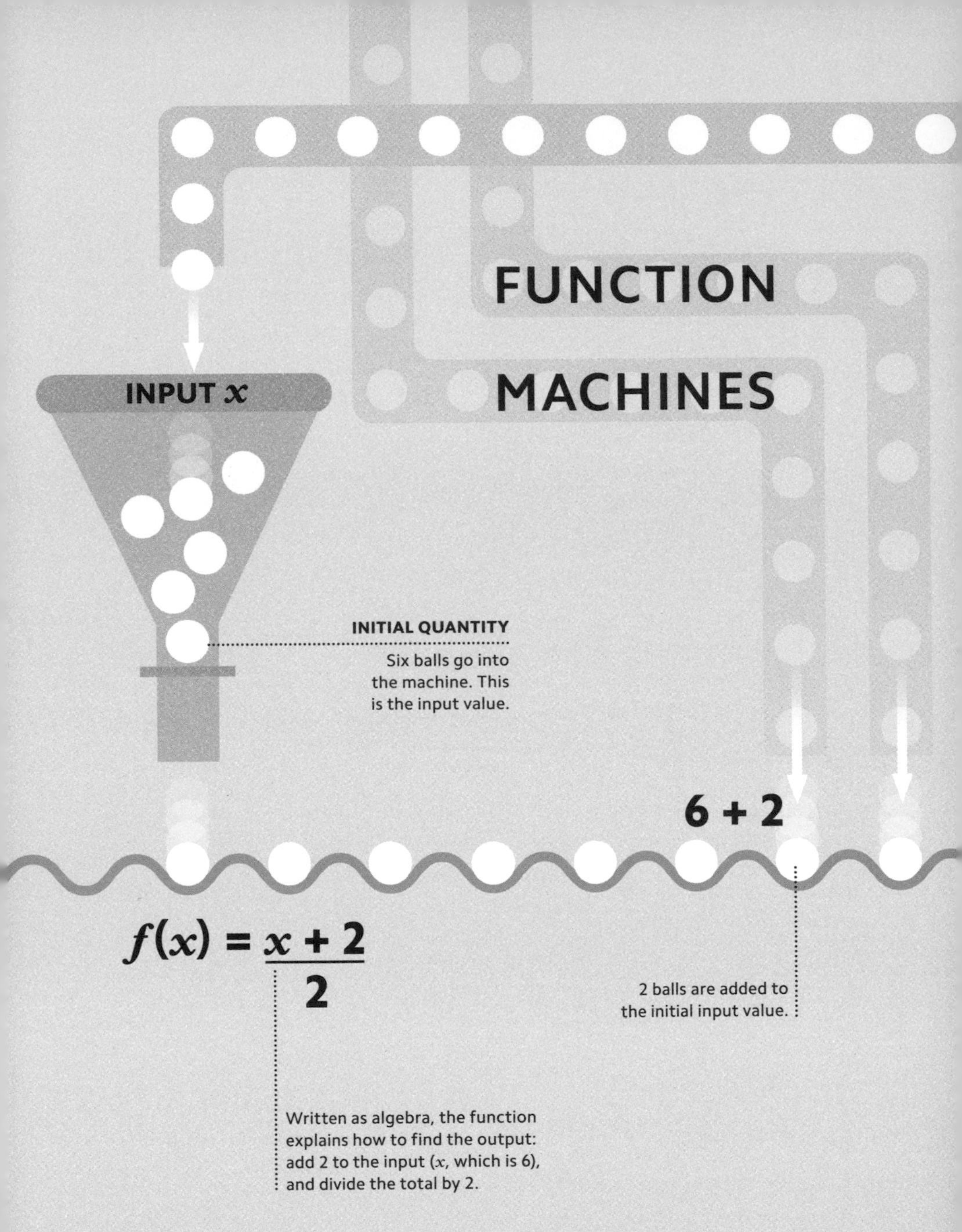

INPUT x

INITIAL QUANTITY
Six balls go into
the machine. This
is the input value.

$6 + 2$

$$f(x) = \frac{x + 2}{2}$$

2 balls are added to
the initial input value.

Written as algebra, the function
explains how to find the output:
add 2 to the input (x, which is 6),
and divide the total by 2.

A function takes an input, processes it according to a specific rule, and creates a unique output. A helpful way to visualize a function is as a machine. In the below example, the input is six, and the output, four. When functions are written as algebra, $f(x)$ represents the function (see below). Although the initial amount of balls in this case is 6, the role of a function is to describe how any input can be transformed. Therefore, the input value is written as x.

÷ 2

The number of balls is halved, making the output value 4.

OUTPUT

How to plot a linear equation

Make a table of values for x and y. Input values for x
into the linear equation to find coordinates. Here, if
$x = 1$, $y = 2x + 1$ becomes $y = (2 \times 1) + 1$, so $y = 3$.

x	-3	-2	-1	0	1	2	3
y	-5	-3	-1	1	3	5	7

$y = 2x + 1$

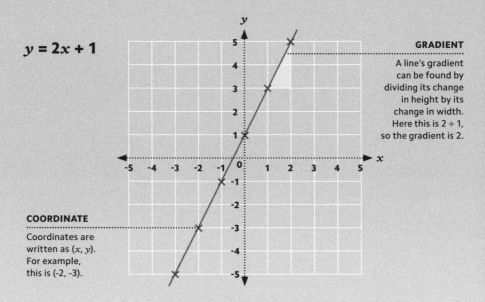

GRADIENT

A line's gradient
can be found by
dividing its change
in height by its
change in width.
Here this is 2 ÷ 1,
so the gradient is 2.

COORDINATE

Coordinates are
written as (x, y).
For example,
this is (-2, -3).

PLOTTING A LINEAR EQUATION

A straight line graph is called a linear graph. It is described by a linear
equation in the form $y = mx + c$, where m is the line's gradient and c is
the y-intercept (the point at which the line intersects the y-axis). When
this is plotted on a grid of coordinates, the result is a graph made up of
a straight line running through all the values of x and y that satisfy it.
For example, for $y = 2x + 1$ the gradient is 2 and the y-intercept is 1.

PARABOLAS

A quadratic graph plots the relationship between x and y for a quadratic equation (see p.72). To draw this graph, the quadratic equation is plotted, creating a U-shaped curve called a parabola. Each parabola has a vertex. This is the lowest (or highest) point on the graph, and the line's turning point. A parabola is symmetrical about a line drawn through its vertex.

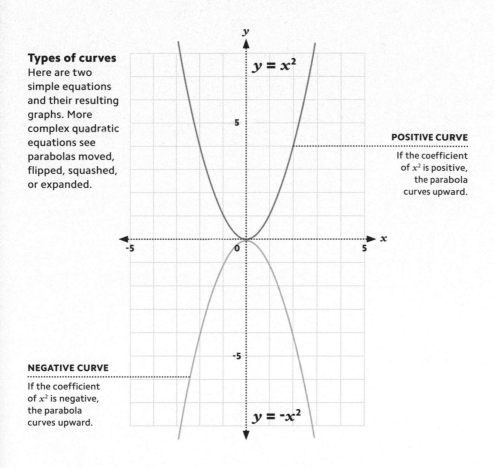

Types of curves
Here are two simple equations and their resulting graphs. More complex quadratic equations see parabolas moved, flipped, squashed, or expanded.

$$y = x^2$$

POSITIVE CURVE

If the coefficient of x^2 is positive, the parabola curves upward.

NEGATIVE CURVE

If the coefficient of x^2 is negative, the parabola curves upward.

$$y = \text{-}x^2$$

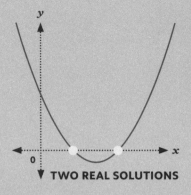

TWO REAL SOLUTIONS

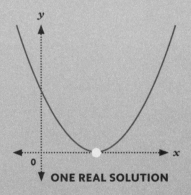

ONE REAL SOLUTION

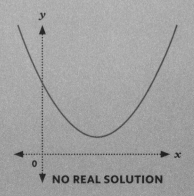

NO REAL SOLUTION

GRAPHING EQUATIONS

All points on the line plotted on a graph represent a solution to the equation that created the line. Linear equations form straight line graphs (see p.82) and quadratic equations form curved parabolic graphs (see p.83). If two linear equations are plotted on the same graph, the point of intersection of the two resulting lines is a solution for both equations. In quadratic graphs (pictured), a quadratic equation in the form of $ax^2 + bx + c = 0$ is plotted (see p.72). This creates a U-shaped line which can be used to determine if there are two real solutions, one real solution, or zero real solutions.

Types of solutions

A quadratic equation can have one of three types of solutions. If one or two real solutions exist, they are found at the points at which the curved line (parabola) intercepts the x-axis.

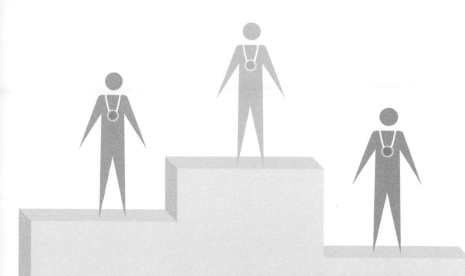

USING REAL-WORLD DATA

Real-life graphs typically display information gathered from observation, or show the relationship between two variables (such as humidity and air temperature). An example of a real-life graph is a distance-time graph (see below). This involves plotting distance against time for a moving object—in this case, three runners in a race. Distance-time graphs allow the speed of each runner to be determined at any point.

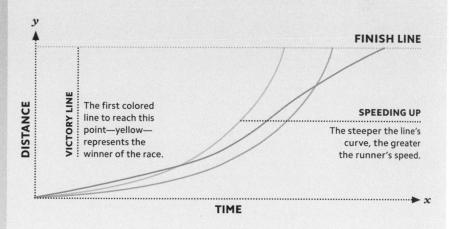

y

DISTANCE

VICTORY LINE

FINISH LINE

The first colored line to reach this point—yellow—represents the winner of the race.

SPEEDING UP

The steeper the line's curve, the greater the runner's speed.

TIME

x

RATIO
PROPO

AND
RTION

Like many ideas in mathematics, the concepts of ratio
and proportion are concerned not just with numbers
themselves, but with the relationships between numbers.
These relationships convey fundamental information
about the world. Ideas of proportion influence not only the
world of facts, but also of aesthetics—with perceptions of
harmonious proportions affecting our judgments of beauty
in art and architecture. More practically, fortunes can be
greatly affected by "compound interest," which causes
an exponential increase over time. Understanding these
relationships helps to model real-world phenomena
and make predictions.

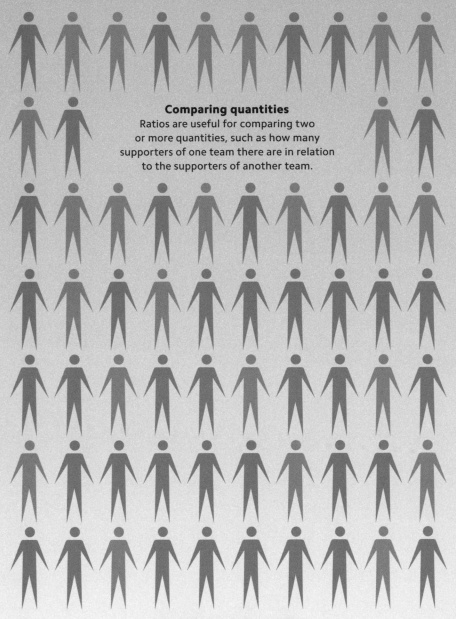

Comparing quantities
Ratios are useful for comparing two or more quantities, such as how many supporters of one team there are in relation to the supporters of another team.

RATIO 24 : 40

Unitary ratios
Currency exchange rates convert
one unit of the base currency into
the equivalent amount of
a different currency.

RATIO 1 : 2

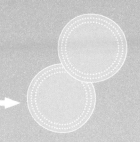

COMPARING AMOUNTS

A ratio is a way of comparing two or more quantities.
It is written $a : b$ to indicate the quantity of a in relation to
the quantity of b. When a ratio has a 1 on one side of it, it's a
unitary ratio. For example, if a fruit bowl has twice as many
apple as oranges, the ratio is 2 : 1. Like fractions, ratios can
be simplified by dividing both sides of the ratio by the
highest common factor—the highest number by
which both ratio numbers can be divided.

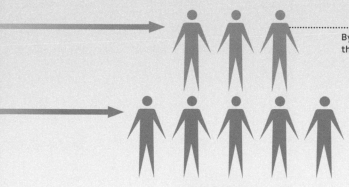

SIMPLIFIED RATIO
By dividing both sides by 8,
the ratio can be reduced to
its simplest form.

RATIO 3 : 5

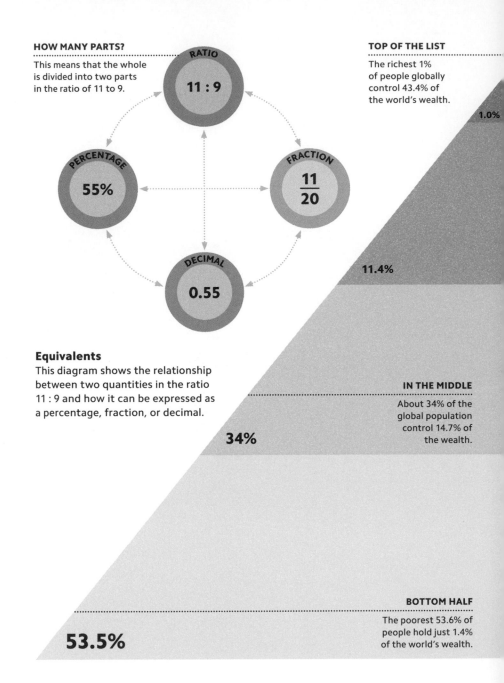

HOW MANY PARTS?

This means that the whole is divided into two parts in the ratio of 11 to 9.

RATIO

11 : 9

TOP OF THE LIST

The richest 1% of people globally control 43.4% of the world's wealth.

1.0%

PERCENTAGE

55%

FRACTION

$$\frac{11}{20}$$

DECIMAL

0.55

11.4%

Equivalents

This diagram shows the relationship between two quantities in the ratio 11 : 9 and how it can be expressed as a percentage, fraction, or decimal.

IN THE MIDDLE

About 34% of the global population control 14.7% of the wealth.

34%

BOTTOM HALF

The poorest 53.6% of people hold just 1.4% of the world's wealth.

53.5%

COMPARISON RELATIONSHIPS

Proportion means the size relationship between different parts of something, or between a part and the whole. Fractions, decimals, ratios, and percentages are all ways of describing proportions—and they are all mathematically interchangeable. Percentages are really a type of fraction, showing the number of parts out of a 100. Percentages also convert easily to decimals: just divide the percentage amount by 100, to find the proportion of one unit (rather than 100 units) that it represents.

Wealth pyramid
Percentages enable complex sets of data to be compared in a clear way. This global wealth pyramid shows that much of the world's wealth is controlled by a disproportionally tiny percentage of individuals.

CALCULATING CHANGE

Percentages provide a clear way of showing changes in quantities—such as a cost increase when tax is added, or a price reduction due to a discount. To calculate a percentage increase, determine the amount of change and add this to the original quantity. A 15% increase means adding $^{15}\!/_{100}$ (or 0.15) times the original quantity. If the total amount after a percentage change is known—such as a price with added tax included—calculating the original amount is called a reverse percentage calculation.

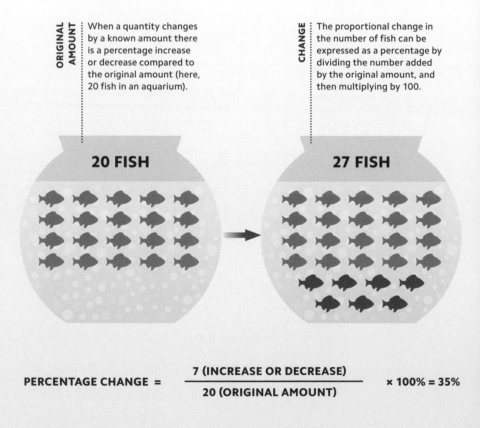

ORIGINAL AMOUNT
When a quantity changes by a known amount there is a percentage increase or decrease compared to the original amount (here, 20 fish in an aquarium).

CHANGE
The proportional change in the number of fish can be expressed as a percentage by dividing the number added by the original amount, and then multiplying by 100.

20 FISH

27 FISH

$$\textbf{PERCENTAGE CHANGE} = \frac{\textbf{7 (INCREASE OR DECREASE)}}{\textbf{20 (ORIGINAL AMOUNT)}} \times \textbf{100\%} = \textbf{35\%}$$

Snowballing interest
This example shows the interest on a capital sum (initial amount) of $1,000 deposited into a savings account that pays 10 percent per annum, compounded monthly.

CAPITAL

$1,000

CAPITAL + INTEREST PAID AFTER 5 YEARS
$1,610.51

CAPITAL + INTEREST PAID AFTER 10 YEARS
$2,593.74

COMPOUND CALCULATION

In this formula, "A" is the final amount after "t" years' interest compounded "n" times, at interest rate "r" on the starting amount of "P".

$$A = P\left(1 + \frac{r}{n}\right)^{nt}$$

CAPITAL + INTEREST PAID AFTER 15 YEARS
$4,177.24

ADDING UP OVER TIME

Savings accounts often pay a small amount of interest, while interest on loans is usually charged at a much higher rate. The longer the loan is held (if none of the loan is paid off) or the savings remain invested, the higher the amount of interest due each month or year. This is called compound interest, as interest is paid not just on the sum invested or borrowed, but also on the interest added to this sum over time.

PROPORTIONAL CHANGES

When things are in proportion to each other—like a model village compared to a real village—this is called direct proportion. In mathematics, direct proportion can be expressed by the equation $y = kx$, where k is the scale factor between x and y. For a model village, for example, k might be 10, meaning that every 1 unit in the model (x) represents 10 units in real life (y). Sometimes quantities are in inverse proportion: as one quantity gets larger, the other gets smaller. This is expressed by the equation $y = {}^k\!/\!x$.

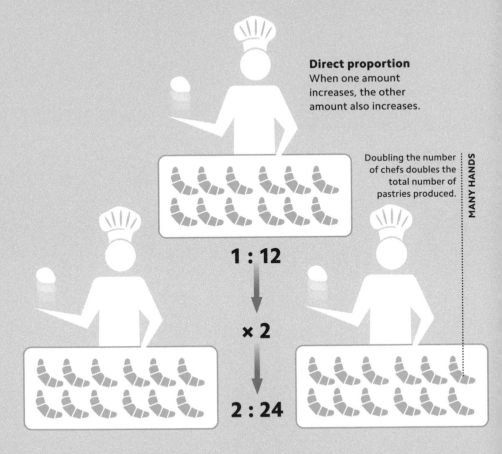

Direct proportion
When one amount increases, the other amount also increases.

Doubling the number of chefs doubles the total number of pastries produced.

MANY HANDS

1 : 12

× 2

2 : 24

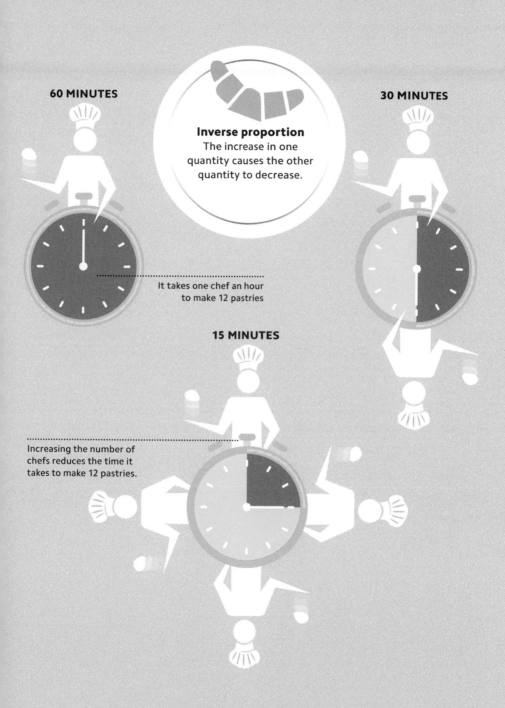

60 MINUTES

30 MINUTES

Inverse proportion
The increase in one quantity causes the other quantity to decrease.

It takes one chef an hour to make 12 pastries

15 MINUTES

Increasing the number of chefs reduces the time it takes to make 12 pastries.

a

Special property
If a square is cut off one end of a rectangle with golden ratio proportions, the rectangle left over also has golden ratio proportions.

a

IDEAL PROPORTIONS

In his *Elements*, the ancient Greek mathematician Euclid described the golden ratio in terms of lengths on a straight line, which is cut in two so that the ratio of the entire line to the longer length is the same as the longer length to the shorter. The golden ratio can be written algebraically as a quadratic equation, which gives only one positive solution—approximately 1.618. The exact number, known as "phi" or the Greek letter Φ, is irrational—it cannot be expressed precisely as a fraction made up of two integers. The golden ratio is closely related to the Fibonacci series (see p.77).

1.618033989...

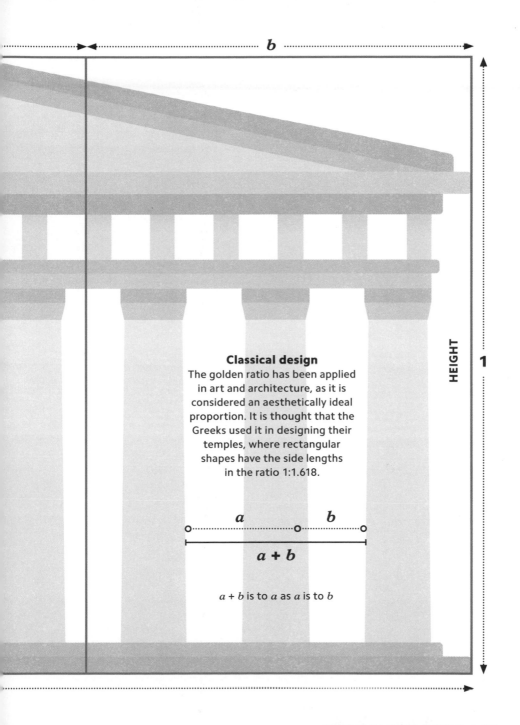

b

HEIGHT

1

Classical design
The golden ratio has been applied in art and architecture, as it is considered an aesthetically ideal proportion. It is thought that the Greeks used it in designing their temples, where rectangular shapes have the side lengths in the ratio 1:1.618.

a   *b*

a + b

a + *b* is to *a* as *a* is to *b*

EXPONENTIAL GROWTH

The phrase "growing exponentially" has a precise mathematical meaning. In an exponential equation, a fixed "base" number is raised to the power of the variable, x. So for the exponential equation $y = 2^x$, the value of y doubles all the time. So the same equation has the capacity to grow slowly at first, then very fast later on—as was seen in the exponential phase of Covid-19 spread. In exponential growth, the more cases there are, the faster their number will increase. But whether numbers are high or low, the time taken for them to double stays the same—like moving one square on the chessboard in the mythical story.

Chess board exponential

In the famous story, an inventor is offered a reward by an Indian emperor, and he chooses rice—starting with one grain of rice on the first square of a chess board, then doubling the number of grains for each subsequent square.

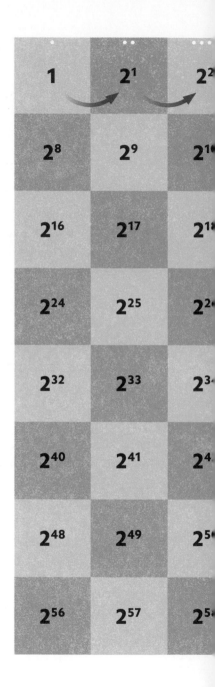

2^3	2^4	2^5	2^6	2^7
2^{11}	2^{12}	2^{13}	2^{14}	2^{15}
2^{19}	2^{20}	2^{21}	2^{22}	2^{23}
2^{27}	2^{28}	2^{29}	2^{30}	2^{31}
2^{35}	2^{36}	2^{37}	2^{38}	2^{39}
2^{43}	2^{44}	2^{45}	2^{46}	2^{47}
2^{51}	2^{52}	2^{53}	2^{54}	2^{55}
2^{59}	2^{60}	2^{61}	2^{62}	2^{63}

SLOW START

On the eighth square, there would be 128 grains of rice, and the total amount so far would be **256 grains**.

HALFWAY

The total number of grains on this square is **2,147,483,648**.

LARGE TOTAL

By the 64th square, the total number of grains on the board is **18,446,744,073, 709,551,616**.

MEAS

U R E

Throughout history, humans have made sense of the physical world by describing it in terms of quantities such as length, mass, and time. Ancient civilizations developed many ways of measuring these quantities, culminating in the standard units we are familiar with today, such as inches, pounds, and seconds. The modern world possesses instruments that can measure these quantities to an extremely high level of accuracy. Standardization creates a common language, while accuracy gives confidence that outcomes are reliable and predictable.

MASS

A range of measures
Different units of
measurement are used to
measure different things:
from the speed of a moving
object to the loudness of
a sound. The latter is
measured in decibels (db).

LOUDNESS

LENGTH, DISTANCE, HEIGHT

SPEED

TIME

TEMPERATURE

CAPACITY

MADE TO MEASURE

A measurement is an attempt to quantify the size of something.
There are two main systems of measurement: imperial and metric.
They use different units of measurement. For example, in imperial,
distances are measured in lengths such as inches (in), feet (ft), and
miles, while in metric, millimeters (mm), centimeters (cm), and
kilometers (km) are preferred. A measurement is only as
accurate as the tool used to make it. Because of this, upper
and lower bounds are useful for describing accuracy (see p.109).

A MATTER OF TIME

Time can be described as the duration of an event or the intervals between events. Standard units of time include seconds, minutes, hours, days, weeks, and years. Early civilizations derived some of these measurements from observing the sun's movement in the sky. This is related to the rotation and orbit of Earth, which takes 24 hours (one day) to rotate on its axis and about 365 days (one year) to orbit the sun.

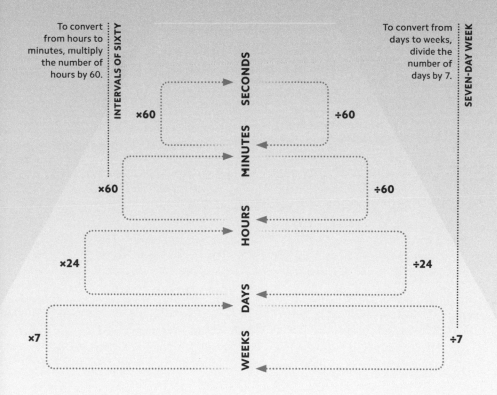

To convert from hours to minutes, multiply the number of hours by 60.

INTERVALS OF SIXTY

To convert from days to weeks, divide the number of days by 7.

SEVEN-DAY WEEK

SECONDS

×60   ÷60

MINUTES

×60   ÷60

HOURS

×24   ÷24

DAYS

×7   ÷7

WEEKS

A parallelogram is made up of two pairs of identical, opposite sides. Adding the length of the base and an adjacent side and doubling the result will give the perimeter.

The formula to find the area of a parallelogram is Area = base × height ($A = b \times h$). For example, with a base of 5 ft and a height of 3 ft, the total area would be 15 ft^2.

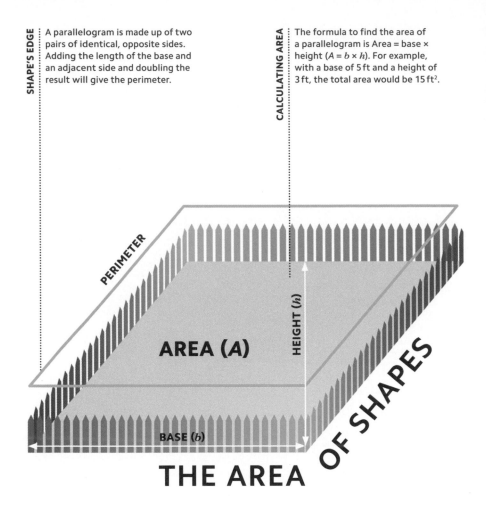

PERIMETER

AREA (*A*)

HEIGHT (*h*)

BASE (*b*)

THE AREA OF SHAPES

Area is the size of a two-dimensional surface. It is measured in square units of length such as square inches (in^2) and square feet (ft^2). Some shapes have a specific formula for calculating their area. A closed shape's perimeter is the sum of the lengths of its sides. Here, the garden is a parallelogram, the size of its surface is its area, and the sum of the garden's edges (the fence) is its perimeter.

TALKING IN CIRCLES

A circle is a round shape that consists of a single closed line made up of points a fixed distance from the circle's center. A circle's circumference (perimeter) divided by the its diameter gives π (pi), which is 3.1416 (to 4 decimal places). As this is true of all circles, π is useful for finding a circle's area or radius when one or the other is unknown. Circles also have unlimited lines of symmetry, and the smallest ratio of perimeter to area of any two-dimensional shape.

SEGMENT

Anatomy of a circle
A circle has many different parts formed by straight lines inside and outside of the circle. Each part has its own specific name.

CHORD
A straight line linking two points on the circumference. It divides a circle into two segments.

CIRCUMFERENCE
The total length of the circle's outer edge.

DIAMETER
A line that cuts a circle exactly in half.

RADIUS
A line from the center to the outer edge. Two radii form a sector between them.

SECTOR

AREA

TANGENT
A line that passes a circle, touching it at one point.

SPACE FILLER

The volume of an object is the amount of three-dimensional space it occupies. The volume of a solid, liquid, or gas is measured in cubic units. An object's volume can also be helpful for calculating its mass. Simple shapes often have a formula that can be used to find their volume. The volume contained within a hollow container is called its capacity.

Cuboid
The cubic units in a cuboid are easy to visualize. To find the shape's volume, multiply its length, width, and height together.
$$V = l \times w \times h$$

Imagine this is a box of sugar cubes. Another way to find the box's volume is to multiply the number of cubes that fit inside by the volume of a single cube.

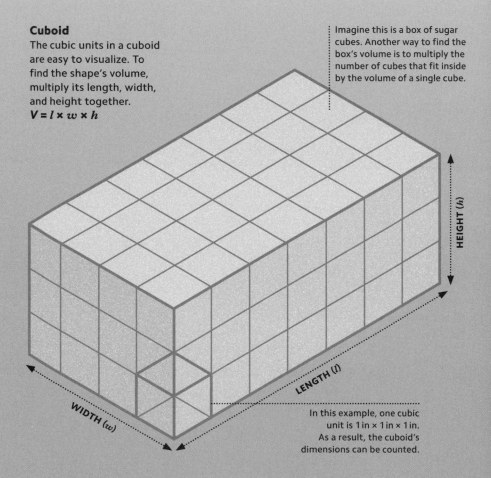

HEIGHT (h)

LENGTH (l)

WIDTH (w)

In this example, one cubic unit is 1 in × 1 in × 1 in. As a result, the cuboid's dimensions can be counted.

Cylinder

A cylinder is made up of two identical and parallel circles connected by a curved surface. Its height is the distance between the circles. A cylinder's radius and height must be known before its volume can be calculated.

$V = \pi \times r^2 \times h$

The radius is the distance from the center of the cylinder's circular face to its outer edge.

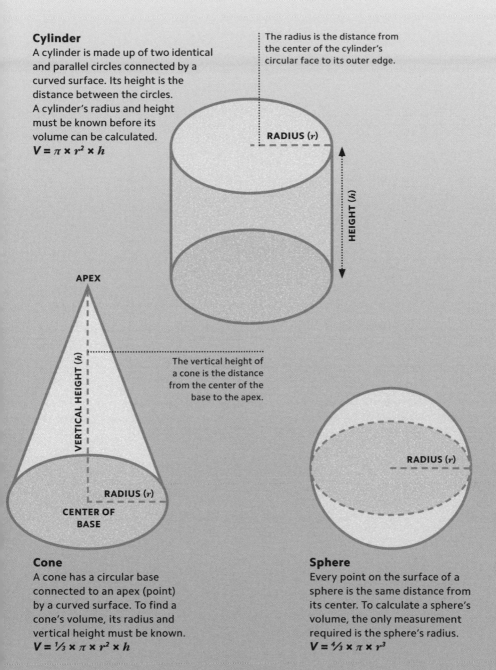

RADIUS (r)

HEIGHT (h)

APEX

VERTICAL HEIGHT (h)

The vertical height of a cone is the distance from the center of the base to the apex.

RADIUS (r)

CENTER OF BASE

RADIUS (r)

Cone

A cone has a circular base connected to an apex (point) by a curved surface. To find a cone's volume, its radius and vertical height must be known.

$V = \tfrac{1}{3} \times \pi \times r^2 \times h$

Sphere

Every point on the surface of a sphere is the same distance from its center. To calculate a sphere's volume, the only measurement required is the sphere's radius.

$V = \tfrac{4}{3} \times \pi \times r^3$

The height of the cylinder in its original state is now the width of this rectangle.

A cylinder's middle section, unrolled, becomes a rectangle.

RADIUS

The circle has the same radius as the cylinder.

1.84 IN

A NET RESULT

8.9 IN

The total area of a three-dimensional object's surface is called its surface area. Knowing an object's surface area has many real-life uses, such as allowing us to find the exact amount of paint needed to cover the object. The surface area of a building or living creature also determines how much heat is lost through its surface. It can be helpful to visualize a shape's surface area as if the shape has been flattened or opened out—this is called its net. Some three-dimensional shapes have their own formula for calculating their surface area.

Opened out flat

This is the net of a cylinder. The cylinder's surface area is the sum of the area of the rectangle and the circles.

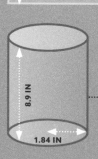

8.9 IN

1.84 IN

ORIGINAL FORM

A cylinder is made up of two identical circles connected by a curved surface.

A DEGREE OF ACCURACY

Defining the bounds
These scales measure an object's weight to the nearest 10 g. As a result, the exact weight of this apple could be anything from 75–85 g. This means the lower bound is 75 g and the upper bound is 85 g.

TO THE NEAREST 10 GRAMS

80g

No measurement can be exact. As a result, there are a number of ways to describe its accuracy. This includes to the nearest amount (such as the nearest 10 g), to the nearest percentage (such as the nearest 5%), or to a certain number of significant figures (e.g. 1.43 miles, to 2 significant figures, is 1.4 miles). The smallest amount that would round to the value of a measurement is called the lower bound, and the top limit is called the upper bound.

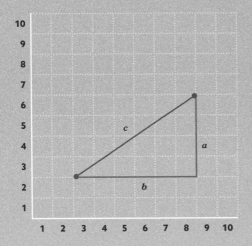

Finding the length

Create a right-angled triangle and write down Pythagoras' Theorem:

$$a^2 + b^2 = c^2$$

Measure lines a and b and input them into the formula:

$$4^2 + 6^2 = c^2$$

Calculate the square numbers:

$$16 + 36 = c^2$$

This gives the length of line c squared:

$$52 = c^2$$

Swap the order, and square root both sides:

$$\sqrt{c^2} = \sqrt{52}$$

This gives the length of line c:

$$c = 7.2 \text{ (to 2 significant figures)}$$

RIGHT-ANGLED TRIANGLES

Pythagoras' theorem derives its name from Pythagoras, the 6th-century-BCE mathematician and philosopher to whom it is traditionally attributed. This theorem allows the length of one side of a right-angled triangle to be found if the lengths of the other two sides are known. In algebra, this is phrased as $a^2 + b^2 = c^2$, where c is the hypotenuse (the side opposite the right angle) and a and b are the other sides.

"Reason is immortal, all else mortal."

Pythagoras

Demonstrating Pythagoras

To demonstrate the theorem, draw the squared sides of the triangle as blocks of squares. Then count the number of individual squares within each block. The number of squares in a^2 and b^2 combined is equal to c^2.

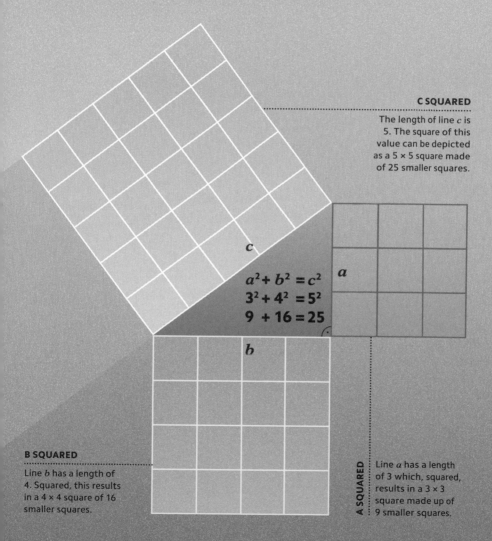

C SQUARED

The length of line c is 5. The square of this value can be depicted as a 5 × 5 square made of 25 smaller squares.

$$a^2 + b^2 = c^2$$
$$3^2 + 4^2 = 5^2$$
$$9 + 16 = 25$$

B SQUARED

Line b has a length of 4. Squared, this results in a 4 × 4 square of 16 smaller squares.

A SQUARED

Line a has a length of 3 which, squared, results in a 3 × 3 square made up of 9 smaller squares.

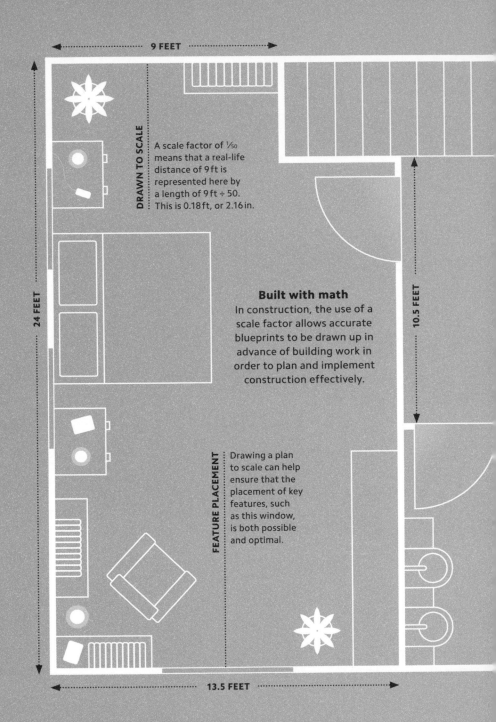

9 FEET

DRAWN TO SCALE
A scale factor of ¹⁄₅₀ means that a real-life distance of 9 ft is represented here by a length of 9 ft ÷ 50. This is 0.18 ft, or 2.16 in.

24 FEET

10.5 FEET

Built with math
In construction, the use of a scale factor allows accurate blueprints to be drawn up in advance of building work in order to plan and implement construction effectively.

FEATURE PLACEMENT
Drawing a plan to scale can help ensure that the placement of key features, such as this window, is both possible and optimal.

13.5 FEET

5.4 FEET

SCALE
FACTOR

Scale drawings and maps share the same proportions as the real-life buildings or areas they depict. They are drawn using a scale factor, which works by multiplying every measurement by the same number. This scale drawing, for example, has a scale factor of $\frac{1}{50}$ (also written as 1:50). This means that every length in the drawing is $\frac{1}{50}$ of the real-life length that it represents. Scale factors have many real-life uses, and they are integral to construction, furniture design, and maps. In 3D, scale factors are used in scale models.

SCALE FACTOR: $\frac{1}{50}$

These dolls are identical in size and shape and are therefore congruent.
IDENTICAL DOLLS

This doll is the same shape as the dolls to its side, but it is a different size. This means they are similar.
SIZE DIFFERENCE

SIMILAR

In geometry, two shapes are referred to as similar if they share the same proportions. For example, if a shape's sides are double the length of another shape's sides, and both shapes have matching angles, they are similar. Furthermore, if two shapes have matching lengths in addition to matching angles they are referred to as congruent. The orientation or position of the shapes does not impact on whether they are similar or congruent. Congruence and similarity can also apply to three-dimensional objects such as dolls (see above).

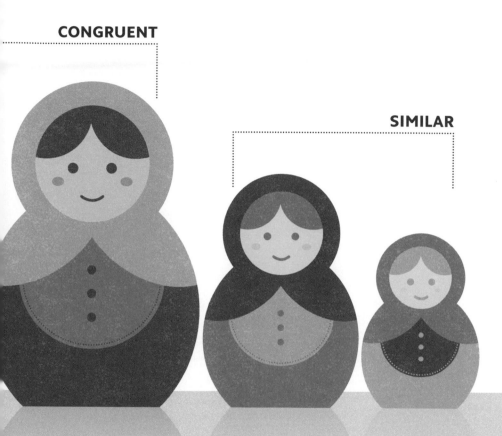

CONGRUENT

SIMILAR

OR THE SAME?

Congruent or similar
Although these dolls share the same
shape, because their proportions are the
same, some are larger than others. All of
the dolls are similar, and the three pairs
of identical size are congruent pairs.

Two congruent shapes
will always share the
same side lengths
and angles.

UNKNOWN ANGLE

To find this angle, the side across from it (the tree's shadow) is labeled the opposite, while the tree's height is the adjacent.

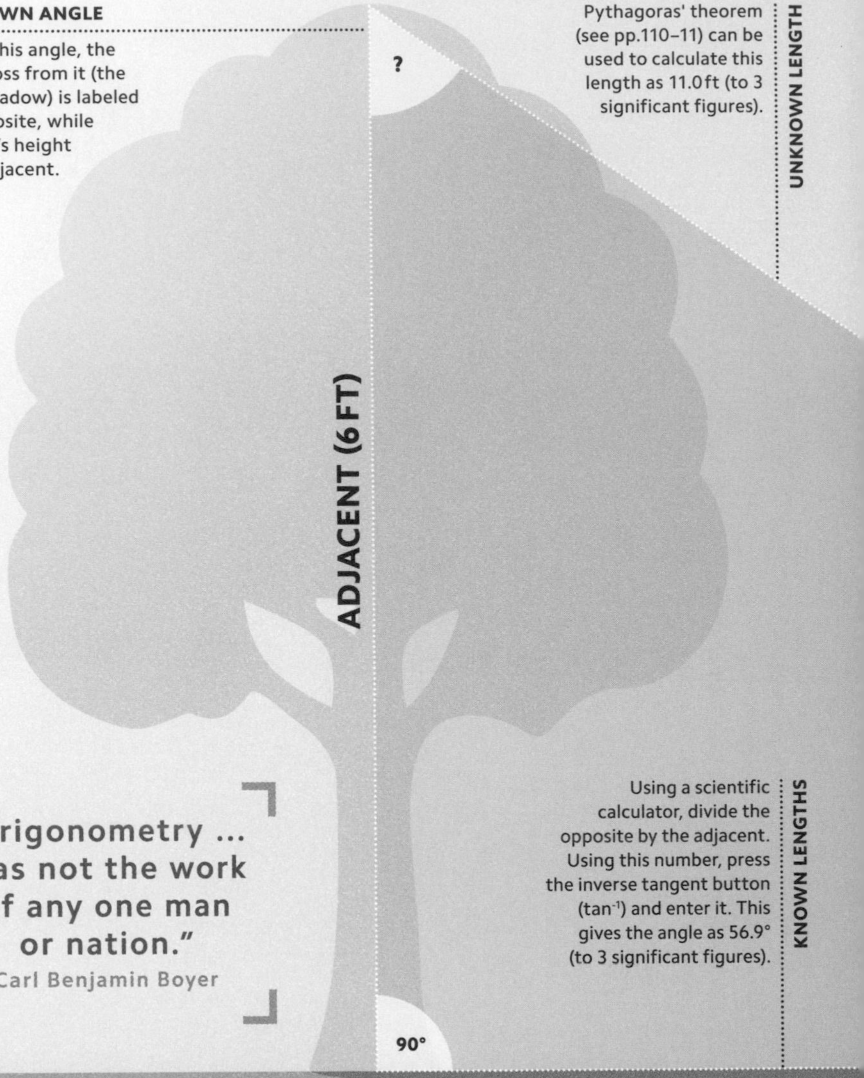

?

UNKNOWN LENGTH

Pythagoras' theorem (see pp.110–11) can be used to calculate this length as 11.0ft (to 3 significant figures).

ADJACENT (6 FT)

"Trigonometry ... was not the work of any one man or nation."
Carl Benjamin Boyer

KNOWN LENGTHS

Using a scientific calculator, divide the opposite by the adjacent. Using this number, press the inverse tangent button (tan⁻¹) and enter it. This gives the angle as 56.9° (to 3 significant figures).

90°

CALCULATING WITH TRIANGLES

Trigonometry is based on the relationships between the side lengths and angles of triangles. In right-angled triangles, the longest side is the hypotenuse. The other two sides, relative to a specific angle, are called the opposite and the adjacent. The three main ratios in trigonometry are sine (sin), cosine (cos), and tangent (tan). The sine is the opposite divided by the hypotenuse, the cosine is the adjacent divided by the hypotenuse, and the tangent is the opposite divided by the adjacent. Because right-angled triangles with the same angles are similar (see pp.114–15) these ratios are always the same.

HYPOTENUSE

Choosing an angle and a ratio

Start by choosing an angle to find—in this case, the top left angle. This angle's opposite and adjacent lengths are both known. Because the tangent is derived from dividing the opposite side length by the adjacent side length, the tangent can be used to find the angle.

?

OPPOSITE (9.2 FT)

The calculations of three or more GPS satellites are so precise that a GPS-enabled smartphone is accurate to within about 16.4 ft under open sky.

DISTANCE OF GPS RECEIVER FROM SATELLITE

The distance calculations of the three satellites intersect at an exact location on Earth's surface. This is where the GPS receiver is located.

GPS RECEIVER

TRIGONOMETRY IN ACTION

Trigonometry is used in many different fields, from construction and navigation to aviation and astronomy. This is because it is useful for calculating distances, heights, and angles. For example, astronomers use trigonometry to measure the distance between stars in the night sky, architects use it to calculate the dimensions of buildings in planning, and Global Positioning Systems (GPS) rely on trigonometry applied to satellite signals to determine a GPS receiver's exact location on Earth.

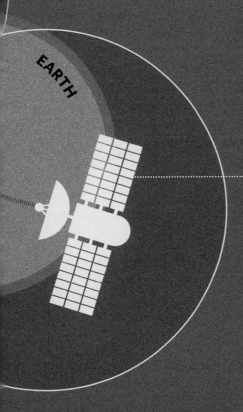

EARTH

SATELLITE
GPS satellites equipped with atomic clocks emit highly precise time stamps so that their location can be pinpointed at all times.

How does GPS work?
A GPS satellite works by giving its position in space and its distance from a GPS receiver on Earth. At least three satellites must do this at once for the intersection of their distance calculations to occur at a single point on Earth's surface. This process is known as trilateration.

OBLIQUE TRIANGLES

An oblique triangle is a triangle without a right angle. The sine and cosine rules are used to find an unknown side or unknown angle of an oblique triangle. They can only be used when some of the triangle's sides or angles are already known. Which rule is used is based on which values are known. When inputting known values into either sine or cosine rule, each angle (for example, A) is paired with its opposite side (a). Sine and cosine are abbreviated on a calculator keypad as sin and cos.

B

a

c

Variations of the rules
There are two versions of the sine rule: one for finding a side (see below) and one for finding an angle. Similarly there are two versions of the cosine rule: one for finding a side and another for finding an angle (see below).

SINE RULE
(UNKNOWN SIDE VERSION)

$$\frac{a}{\sin A} = \frac{b}{\sin B} = \frac{c}{\sin C}$$

COSINE RULE
(UNKNOWN ANGLE VERSION)

$$\cos A = \frac{b^2 + c^2 - a^2}{2bc}$$

C

A

b

The speed of a moving object can be found by dividing the amount of time it has been moving by the distance it has traveled.

The distance an object has traveled can be found by multiplying its speed by the amount of time it has been moving.

DISTANCE

DISTANCE = SPEED × TIME

SPEED | TIME

SPEED = DISTANCE ÷ TIME   **TIME = DISTANCE ÷ SPEED**

COMBINING UNITS

A compound measure uses two or more different units. For example, the speed of a moving object is often measured in miles per hour (mph) or kilometres per hour (kph). These compound measures account for both the amount of distance traveled and the amount of time that passes. Other compound measures include pressure, force, density, and the area and volume of objects (see pp.106–107).

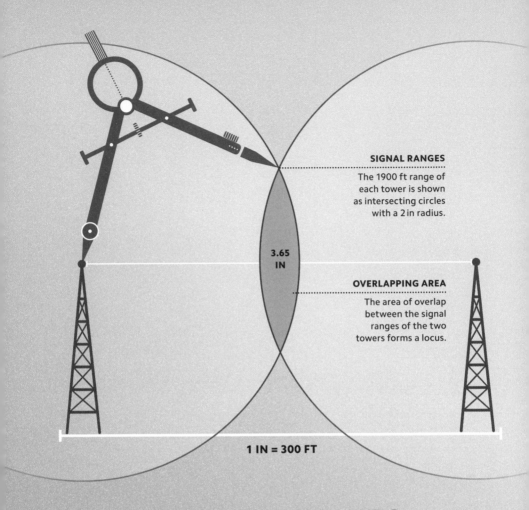

SIGNAL RANGES

The 1900 ft range of each tower is shown as intersecting circles with a 2 in radius.

3.65 IN

OVERLAPPING AREA

The area of overlap between the signal ranges of the two towers forms a locus.

1 IN = 300 FT

SETS OF POINTS

A construction is an accurate drawing made with a pair of compasses and a straight edge (usually a ruler). This could be a drawing of a locus (plural: loci)—a set of points which meet one or several conditions. One example of a locus is a circle (see opposite). Loci can be in 1D, 2D, or 3D. Loci can be combined to form a new locus, such as in this simplified scale drawing of two transmission towers (see above). It shows a locus formed by the area of overlap between two loci (circles).

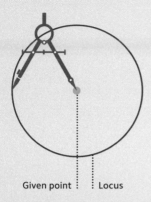

Given point ┊ ┊ Locus

Set distance from a point
Set a pair of compasses at a given point and draw a full circle. This locus meets the condition of its points all being situated the same distance from its center.

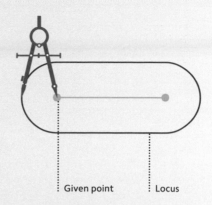

Given point ┊ ┊ Locus

Set distance from a line segment
This locus consists of points a set distance from a line segment. Draw the semicircular ends with a pair of compasses, and then the straight sides with a ruler.

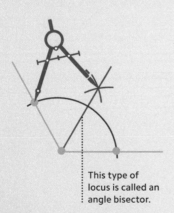

This type of locus is called an angle bisector.

Equidistant from line segments
This locus is a straight line made up of a set of points that meet the condition of lying equidistant (the same distance) from the two intersecting yellow line segments.

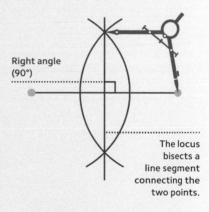

Right angle (90°)

The locus bisects a line segment connecting the two points.

Equidistant from two points
Drawing a locus equidistant from two points forms a perpendicular bisector: a straight line that passes (at 90°) through the midpoint of a line segment joining the two points.

In a still body of water, the boat's route would be a vector of dimensions 30 across and 0 up or down.

INTENDED DIRECTION OF TRAVEL

ACTUAL DIRECTION OF TRAVEL

MAGNITUDE AND DIRECTION

A vector is a distance in a certain direction. Every vector has a size and a direction. In diagrams this is drawn as a line with an arrow on it. The length of the line indicates the size of the vector while the arrow signifies the direction. A vector is often expressed by giving the horizontal units (for example, yards) over the vertical units (see opposite). If the horizontal units are positive, the direction of the vector is right, and if negative, left. Similarly, if the vertical units are positive, the vector moves up, and if negative, down.

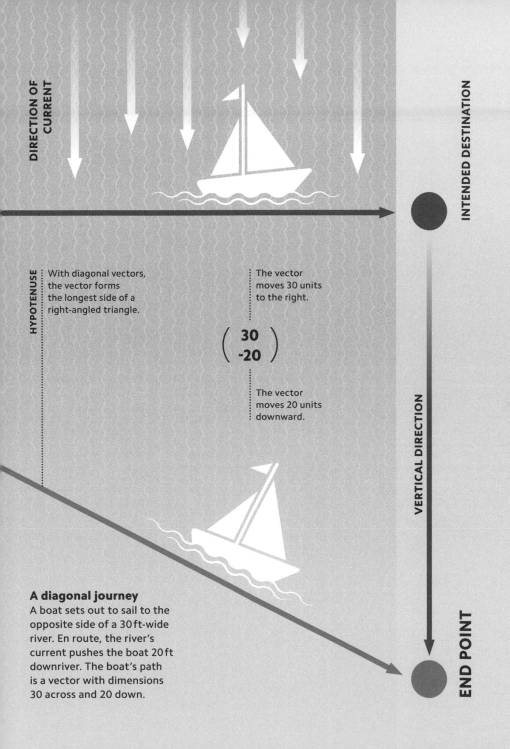

DIRECTION OF CURRENT

INTENDED DESTINATION

HYPOTENUSE

With diagonal vectors, the vector forms the longest side of a right-angled triangle.

The vector moves 30 units to the right.

$$\begin{pmatrix} 30 \\ -20 \end{pmatrix}$$

The vector moves 20 units downward.

VERTICAL DIRECTION

END POINT

A diagonal journey
A boat sets out to sail to the opposite side of a 30 ft-wide river. En route, the river's current pushes the boat 20 ft downriver. The boat's path is a vector with dimensions 30 across and 20 down.

WHAT IS THE MATRIX?

A matrix (plural: matrices) is a rectangular grid of numbers or variables enclosed by square or curved brackets. Matrices record information in rows and columns that can be extended to hold huge amounts of data. They allow a computer to automate calculations and have applications in areas such as physics, computer science, and cryptography. Matrices can be added, subtracted, or multiplied. They are also used to transform two-dimensional shapes (see opposite) and three-dimensional shapes.

3 + 4 = 7

Manipulating matrices
To add or subtract matrices, each element
in the first matrix is added to or subtracted from
the corresponding element in the second matrix.
To multiply matrices, each row in the first matrix is
multiplied by each column in the second matrix.

The first verified usage of
matrices occurred in ancient
China in the 2nd century BCE.

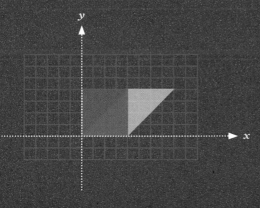

Horizontal shear
This matrix applies a horizontal shear to the original square (purple), slanting it to the right.

$$\begin{bmatrix} 1 & 1 \\ 0 & 1 \end{bmatrix} \times \begin{bmatrix} x \\ y \end{bmatrix}$$

Reflection
This matrix reflects the original square (purple) across a line of symmetry at the y-axis.

$$\begin{bmatrix} -1 & 0 \\ 0 & 1 \end{bmatrix} \times \begin{bmatrix} x \\ y \end{bmatrix}$$

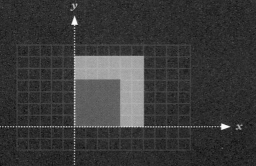

Enlargement
This matrix enlarges the original square (purple) by factor 1.5.

$$\begin{bmatrix} 1.5 & 0 \\ 0 & 1.5 \end{bmatrix} \times \begin{bmatrix} x \\ y \end{bmatrix}$$

STATI
AND
PROBA

STICS
BILITY

Probability and statistics both deal with data. Statistics takes existing data and analyzes and presents the information to uncover trends or relationships between things. Governments need statistics to plan effectively, and thus spend a lot of time and effort in collecting data. Probability was born out of gambling and insurance, where it is profitable to understand the chance of specific outcomes happening. In probability, data is used to predict the chance of future events. Probability is used in all areas of life, including insurance and risk analysis, as well as quantum physics.

ARE STATISTICS VITAL?

Data is information and can be quantitative (involving numbers) or qualitative (involving opinions or descriptions). Statistics is concerned with the collection, organization, presentation, analysis, and interpretation of numerical data. The collected information is used to answer questions or solve problems. Scientific, medical, industrial, or social issues are all addressed using statistics. Mathematics is essential to present and analyze numerical data in effective ways.

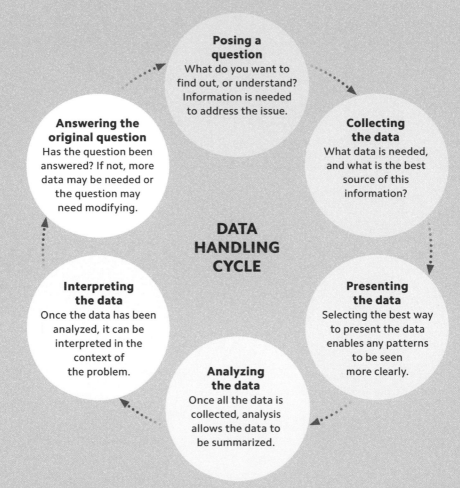

Posing a question
What do you want to find out, or understand? Information is needed to address the issue.

Answering the original question
Has the question been answered? If not, more data may be needed or the question may need modifying.

Collecting the data
What data is needed, and what is the best source of this information?

DATA HANDLING CYCLE

Interpreting the data
Once the data has been analyzed, it can be interpreted in the context of the problem.

Presenting the data
Selecting the best way to present the data enables any patterns to be seen more clearly.

Analyzing the data
Once all the data is collected, analysis allows the data to be summarized.

SURVEYS AND SAMPLES

Data is collected within a group, which is called a population by statisticians. Information can be collected from a whole population, for example in a census questionnaire. Sampling is used to gather information from a subset of the population, with the assumption the data will apply to the whole. Surveys of a randomly selected subset achieve this, but care is needed to avoid potential bias.

Data collection

Good quality information is highly valued, and a lot of time and money is spent by governments and companies to collect data.

DATA CHARTS

Statistics is all about handling data to gain insight into an idea or theory. The data captured is collected in the form of numbers or words. Presenting data in a visual form enables clear understanding of any patterns and trends and comparison of different data sets. The diagrams used depend on the type of data. Data can have discrete values, such as the number of workers in a company, or it can be nonnumerical (qualitative, or categorical), such as favorite type of food.

CATEGORY	FREQUENCY
ORANGE	3
GREEN	7
PURPLE	6
RED	3
YELLOW	4

Table of data
A table is the simplest way to organize data. Frequency counts how often a value occurs in a category.

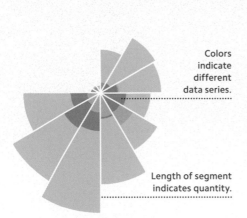

Coxcomb diagram
Arranged like pie charts, coxcomb diagrams are useful for data with cyclical values, such as months of the year.

Colors indicate different data series.

Length of segment indicates quantity.

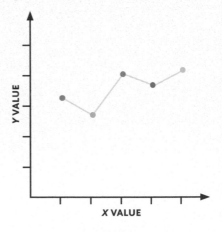

Line graph
Line graphs are typically used for continuous quantitative data. Trends over time are commonly presented in this way.

Y VALUE

X VALUE

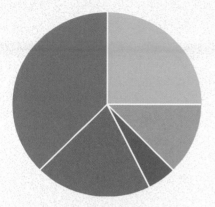

Pie chart
This shows data as a proportion of a whole, showing the fraction of data falling into each category.

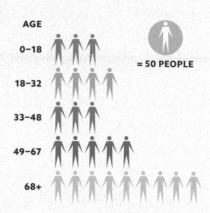

AGE

0–18

18–32

33–48

49–67

68+

= 50 PEOPLE

Pictogram
Pictograms are similar to bar charts but they represent frequency with pictures in place of the bars.

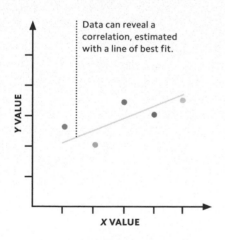

Data can reveal a correlation, estimated with a line of best fit.

Y VALUE

X VALUE

Scatter graph
Plotting two continuous variables against each other can show if there is a correlation between them (for example, age and height).

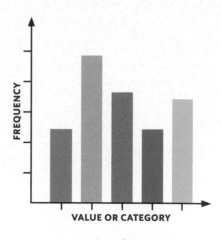

FREQUENCY

VALUE OR CATEGORY

Bar chart
Bar charts are useful for displaying categorical or discrete data, and give the frequency of each category or value.

MAKING SENSE OF DATA

The modern world generates data at an ever-growing rate, and statistics can make sense of it. One key attribute of a data set is average, which gives an idea of a typical value. Another attribute is spread, which indicates how variable the data is. For example, a data set could be analyzed to find the average length of throw of a shot-putter and how consistent they are. Statistics allows analysis of patterns and trends, or possible correlation between two factors, such as CO_2 concentration and global temperature. By analyzing data, informed decisions can be made by governments and businesses.

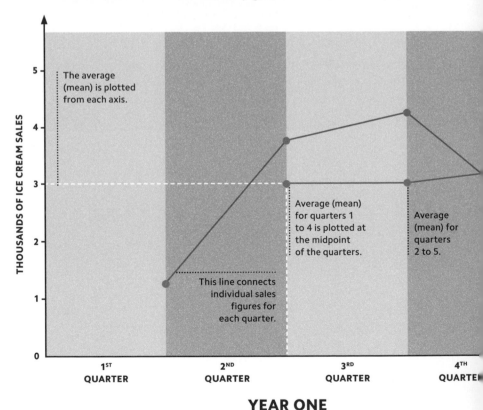

The average (mean) is plotted from each axis.

Average (mean) for quarters 1 to 4 is plotted at the midpoint of the quarters.

Average (mean) for quarters 2 to 5.

This line connects individual sales figures for each quarter.

THOUSANDS OF ICE CREAM SALES

1ST QUARTER   2ND QUARTER   3RD QUARTER   4TH QUARTER

YEAR ONE

Sales data

By analyzing sales, businesses can make decisions, such as stock levels and marketing spend. The table below shows sales of ice cream over a two-year period, with each year divided into four quarters.

	YEAR ONE				YEAR TWO			
QUARTER	1ST	2ND	3RD	4TH	5TH	6TH	7TH	8TH
SALES IN THOUSANDS	1.25	3.75	4.25	2.5	1.5	4.75	5.0	2.75

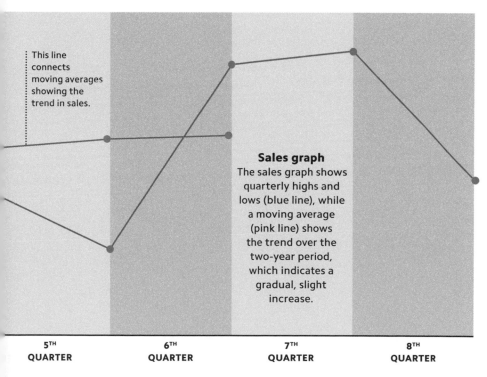

This line connects moving averages showing the trend in sales.

Sales graph

The sales graph shows quarterly highs and lows (blue line), while a moving average (pink line) shows the trend over the two-year period, which indicates a gradual, slight increase.

| 5TH QUARTER | 6TH QUARTER | 7TH QUARTER | 8TH QUARTER |

YEAR TWO

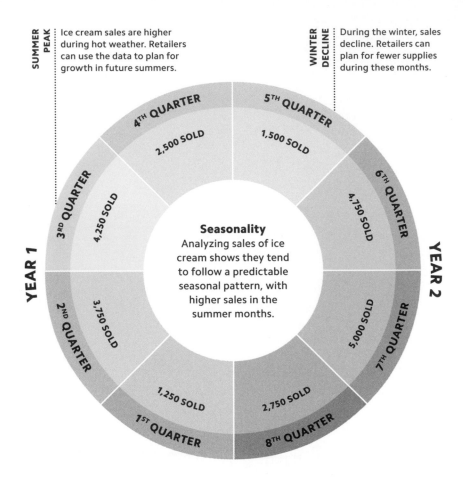

Ice cream sales are higher during hot weather. Retailers can use the data to plan for growth in future summers.

During the winter, sales decline. Retailers can plan for fewer supplies during these months.

4TH QUARTER
2,500 SOLD

5TH QUARTER
1,500 SOLD

3RD QUARTER
4,250 SOLD

6TH QUARTER
4,750 SOLD

YEAR 1

YEAR 2

Seasonality
Analyzing sales of ice cream shows they tend to follow a predictable seasonal pattern, with higher sales in the summer months.

2ND QUARTER
3,750 SOLD

7TH QUARTER
5,000 SOLD

1ST QUARTER
1,250 SOLD

8TH QUARTER
2,750 SOLD

WHAT DOES IT MEAN?

Data interpretation is the process of saying what the analysis of the data means in the context of the problem. Clinical trials test a hypothesis ("this drug is effective against this disease"), and interpretation of the data determines the degree to which the hypothesis is true. Monitored data, such as electicity usage or annual sales, is analyzed to predict requirements and informs strategy.

FINDING CONNECTIONS

Scatter graphs allow us to display data drawn from two different quantities, to see if they might be connected, or correlated. These are known as scatter graphs, and if the data is linked there could be a positive correlation (they increase together), or a negative correlation (one increases as the other decreases). A line of best fit can also be drawn and used for prediction, although this is only reliable within the range of the data. If there is no correlation, then the variables are independent of each other.

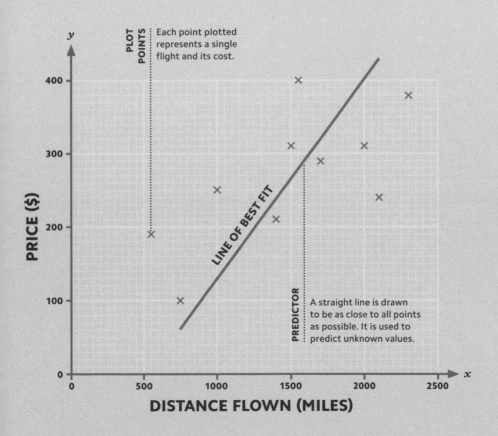

PLOT POINTS
Each point plotted represents a single flight and its cost.

LINE OF BEST FIT

PREDICTOR
A straight line is drawn to be as close to all points as possible. It is used to predict unknown values.

Axis labels: PRICE ($) (y-axis: 0, 100, 200, 300, 400); DISTANCE FLOWN (MILES) (x-axis: 0, 500, 1000, 1500, 2000, 2500)

MEASURING CHANCE

What are the chances of that happening? This is a question that relies on probability to answer. The chance of any event occurring is represented on a scale of probabilities ranging from no chance (0) to certainty (1). Games of chance were the inspiration to develop ideas of probability. These ideas have been developed into examples we encounter in everyday life: insurance, weather prediction, political polling, and clinical trials.

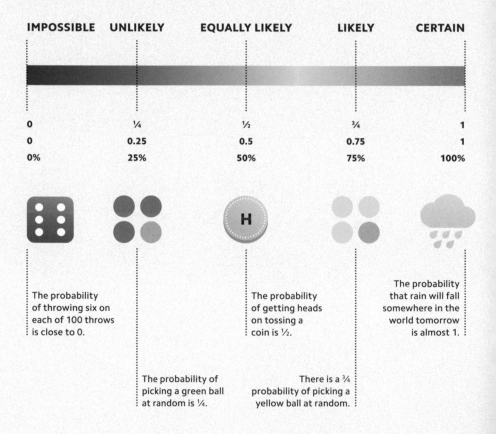

IMPOSSIBLE	UNLIKELY	EQUALLY LIKELY	LIKELY	CERTAIN
0	¼	½	¾	1
0	0.25	0.5	0.75	1
0%	25%	50%	75%	100%

The probability of throwing six on each of 100 throws is close to 0.

The probability of picking a green ball at random is ¼.

The probability of getting heads on tossing a coin is ½.

There is a ¾ probability of picking a yellow ball at random.

The probability that rain will fall somewhere in the world tomorrow is almost 1.

COUNTING OUTCOMES

People value certainty—they want to know the likelihood of an event occurring. Very often, these events can occur in a number of ways. The event "being late for work" could depend on two outcomes: "waking up late" and "traffic bad." Throwing two dice gives 36 possible outcomes. The outcomes can be listed in a table.

Calculating chance

Knowing the number of outcomes leading to an event, such as getting a particular dice score, enables us to calculate the probability.

	●	●●	●●●	●●●●	●●●●●	●●●●●●
●	2	3	4	5	6	7
●●	3	4	5	6	7	8
●●●	4	5	6	7	8	9
●●●●	5	6	7	8	9	10
●●●●●	6	7	8	9	10	11
●●●●●●	7	8	9	10	11	12

MORE LIKELY

There are six ways to throw a seven with two dice.

There is only one way to throw a 12 with two dice.

LESS LIKELY

PROBABILITY OF AN EVENT $=\dfrac{\text{NUMBER OF WAYS THE EVENT CAN HAPPEN}}{\text{TOTAL NUMBER OF POSSIBLE OUTCOMES}}$

Wheel of chance
There is only one way for the spinner to land on a 2, but there are five possible outcomes. So the theoretical probability of landing on a 2 is ⅕.

IN A PERFECT WORLD

Theoretical probability is the chance we assign to an event happening, based on an understanding of the list of possible outcomes. For example, a dice has six possible outcomes (1 to 6). These outcomes are mutually exclusive—they can't occur simultaneously. The event "rolling a multiple of 3" has two outcomes: 3 and 6. We can use the formula at the top of the page to find the probability of this event, which is ⅔ or ⅓.

TESTING THE ODDS

Theoretical probability predictions can be tested experimentally by repeating the trial—throwing dice, for example—and recording the outcomes. The number of events recorded is important. The more observed, the closer to theoretical probability the result should be. In the real world, experiments are used to generate data. The more information gathered, the better the predictions become. Clinical trials, for example, are conducted on drugs to confirm that a treatment works.

EXPERIMENTAL PROBABILITY

=

NUMBER OF TIMES THE EVENT OCCURS

NUMBER OF TRIALS

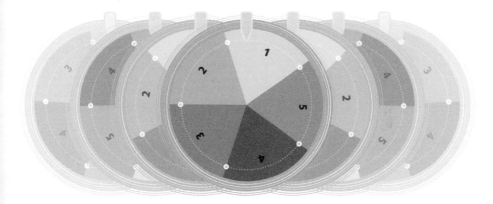

Spinner odds

In theory, the chances of spinning a 1 with this spinner is $\frac{1}{5}$ (20%). The 50 spins suggest the actual probability could be $\frac{6}{50}$ (12%). A lot more spins are needed to confirm whether the spinner is actually biased slightly against 1.

NUMBER	OCCURRENCES
1	6
2	11
3	10
4	14
5	9
TOTAL	**50 TRIALS**

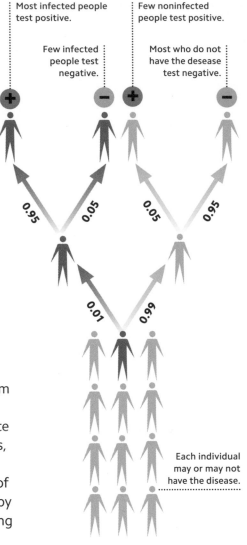

Infection rates

This tree maps the effectiveness of a disease test. It shows the test is correct 95% of the time.

Most infected people test positive.

Few noninfected people test positive.

Few infected people test negative.

Most who do not have the desease test negative.

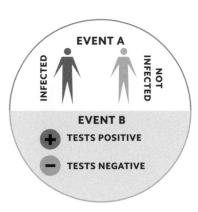

EVENT A

INFECTED

NOT INFECTED

EVENT B

+ TESTS POSITIVE

− TESTS NEGATIVE

0.95

0.05

0.05

0.95

0.01

0.99

Each individual may or may not have the disease.

It is possible to calculate the chances of two independent events happening. A tree diagram can be used to demonstrate all possible outcomes. Each separate event is represented as branches, with the number or probability indicated. For any combination of events, the probability is found by multiplying the probabilities along those branches. In the example shown, for instance, not having the disease and testing positive (false positive) is 0.99 × 0.05 = 0.0495, which is nearly a 1 in 20 chance.

TREE OF POSSIBILITY

PART OF THE UNION

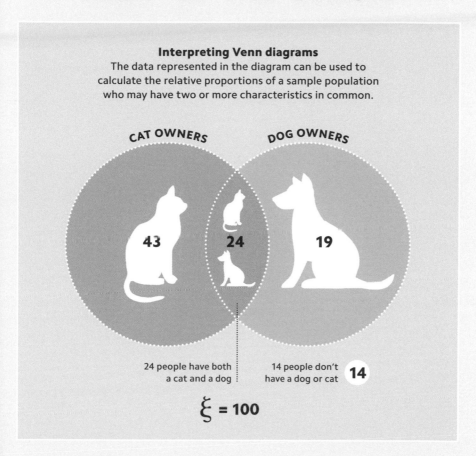

Interpreting Venn diagrams
The data represented in the diagram can be used to calculate the relative proportions of a sample population who may have two or more characteristics in common.

CAT OWNERS

DOG OWNERS

43

24

19

24 people have both
a cat and a dog

14 people don't
have a dog or cat

14

$\xi = 100$

Venn diagrams are a familiar pictorial representation of sets of data, or probabilities, and how they overlap. A Venn diagram has one or more overlapping circles inside a rectangle. The rectangle represents the universal set (all possible outcomes). The circles represent events, or a subset of the data. The information in each region can show merging events and outcomes, the number of event outcomes, or the probability of the events.

How many?

The overlap in a Venn diagram indicates how likely it is that two events happen together.

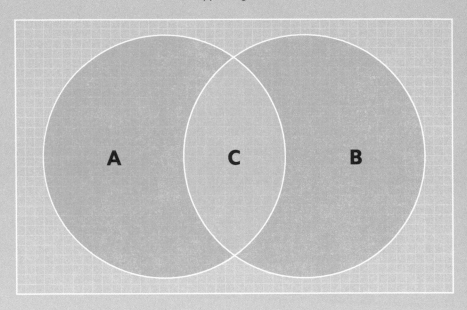

● A = OVER 30 YEARS OLD   ● B = WEARS GLASSES   ● C = OVER 30 AND WEARS GLASSES

IT DEPENDS

Conditional probability is the chance of one event happening based on a known previous event. For example, knowing somebody's age means predictions about them having a specific medical condition are more accurate than assuming they are typical of the entire population. Bayes' theorem is a formula that describes this, and is used to determine important probabilities, with sometimes surprising results. Venn diagrams can be used to organize this information and illustrate conditional probability problems.

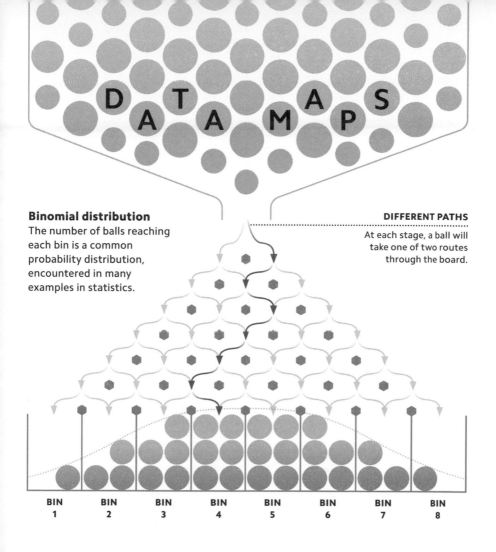

Binomial distribution

The number of balls reaching each bin is a common probability distribution, encountered in many examples in statistics.

DIFFERENT PATHS

At each stage, a ball will take one of two routes through the board.

| BIN 1 | BIN 2 | BIN 3 | BIN 4 | BIN 5 | BIN 6 | BIN 7 | BIN 8 |

Probability distributions map the likelihood of an event. For example, recording the heights of subjects in a random sample creates a distribution of heights. This can be recorded on a normal distribution graph (or bell curve), where the data groups around a central mean (average), with values tailing off either side. A uniform distribution is created where outcomes are equally likely, such as throwing dice. Binomial distributions are created when events have only two outcomes and are repeated multiple times.

VERY UNLIKELY EVENTS

The law of large numbers describes how experimental probability approaches theoretical probability as more events are observed (see p.141). If the number of repetitions of the experiment is infinite, then all possibilities, even outrageous ones, are valid. One famous illustration of the idea is that a monkey, typing randomly and given an infinite amount of time, would eventually write Shakespeare's plays. This shows that high improbability is the same as impossible.

> **University of Plymouth researchers gave monkeys a typewriter for a month. The primates only filled five pages of text, composed primarily of the letter S.**

Infinite time

More than just a very long time, infinity assumes that the universe will continue forever and all possibilities of a recurring event will happen.

CALC

U L U S

Calculus is a branch of mathematics developed independently by German mathematician Gottfried Leibniz and English scientist Isaac Newton. It answers two fundamental questions: how to determine rates of change and how to calculate the area under a curve. Differential calculus is concerned with the rate of change and integral calculus with the sum of a function, or the area under a curve. Integration and differentiation are inverse operations, in other words, they undo one another.

MEASURING CHANGE

The rate at which something changes over time can provide valuable information, such as how fast something is traveling, how quickly an infection is spreading, or whether or not the economy is stagnating. A graph of the varying quantity is likely to be a curve, and the rate of change at any instant can be estimated by drawing a tangent to the curve (a straight line that just touches the curve at that point) and determining the gradient of the tangent.

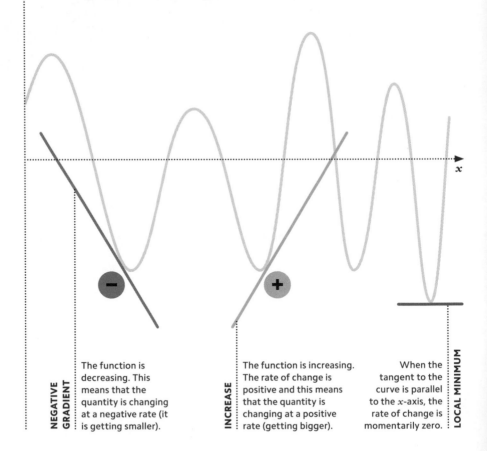

NEGATIVE GRADIENT
The function is decreasing. This means that the quantity is changing at a negative rate (it is getting smaller).

INCREASE
The function is increasing. The rate of change is positive and this means that the quantity is changing at a positive rate (getting bigger).

LOCAL MINIMUM
When the tangent to the curve is parallel to the x-axis, the rate of change is momentarily zero.

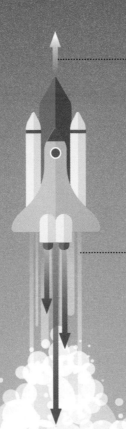

Rocket science

Being able to calculate the rates of change is important in aerospace. For example, as a rocket accelarates away from Earth, it burns fuel, which decreases its mass and increases it acceleration.

MOMENTUM

To overcome the pull of gravity, a rocket needs sufficient momentum.

THRUST

In order for the rocket to lift off, a large amount of thrust is needed. Thrust can be defined as the rate of change (the derivative) of momentum.

CHANGING VALUES

The derivative is a measure of the rate of change of a quantity. The graphs of phenomena that change (such as infections rates, population growth, or financial quantities) can be modeled by a function (see pp.80–81) relating two quantities, one of which is often time. The derivative defines how one variable changes as the other (for example, time) changes. The rate of change at any point can be estimated by determining the gradient of the tangent (see opposite). When this is done algebraically, the derivative function is obtained.

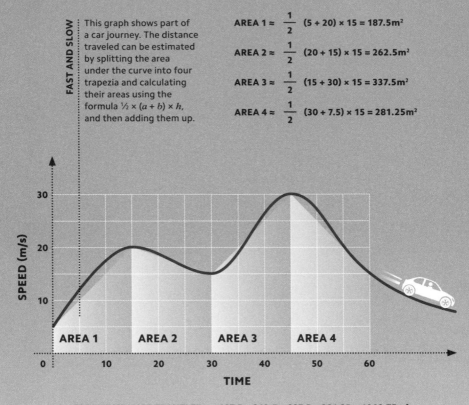

FAST AND SLOW

This graph shows part of a car journey. The distance traveled can be estimated by splitting the area under the curve into four trapezia and calculating their areas using the formula ½ × (a + b) × h, and then adding them up.

AREA 1 ≈ $\frac{1}{2}$ (5 + 20) × 15 = 187.5m²

AREA 2 ≈ $\frac{1}{2}$ (20 + 15) × 15 = 262.5m²

AREA 3 ≈ $\frac{1}{2}$ (15 + 30) × 15 = 337.5m²

AREA 4 ≈ $\frac{1}{2}$ (30 + 7.5) × 15 = 281.25m²

SPEED (m/s)

30

20

10

0   10   20   30   40   50   60

AREA 1   AREA 2   AREA 3   AREA 4

TIME

TOTAL DISTANCE TRAVELED ≈ 187.5 + 262.5 + 337.5 + 281.25 ≈ 1068.75m²

SUMMATION OF QUANTITIES

Finding the area under a curve is a common problem in mathematics. For example, to find the distance traveled from a speed-time graph, you simply need to find the area under the curve. The simplest situation is to find the area under a straight line. However, this idea can be extended to estimate the area under any curve, by splitting the area into a series of strips of equal width and estimating the area of each strip by approximating it to a trapezium (see p.46). The greater the number of strips, the better the estimate.

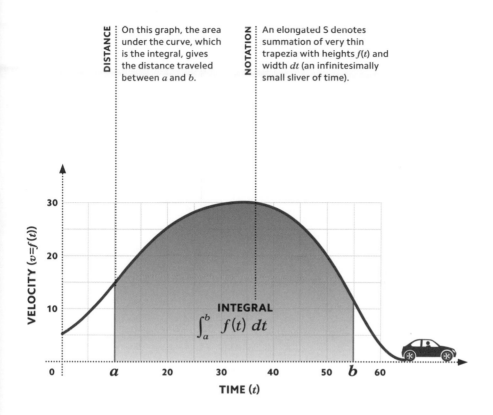

DISTANCE On this graph, the area under the curve, which is the integral, gives the distance traveled between a and b.

NOTATION An elongated S denotes summation of very thin trapezia with heights $f(t)$ and width dt (an infinitesimally small sliver of time).

INTEGRAL

$$\int_a^b f(t)\,dt$$

VELOCITY ($v=f(t)$)

TIME (t)

30

20

10

0   a 20   30   40   50 b 60

AREA UNDER A CURVE

Integration is the algebraic extension of the summation method (see opposite). Imagine the strips getting so thin that they melted together. Using algebra, you can express this mathematically to determine the exact area under a curve, by "integrating" the curve's function. Integration and differentiation (finding the derivative) are inverse processes (see p.154), and the relationship between the two is the basis behind the field known as calculus.

Differentiate
Differentiating a curve's equation for a specified point gives the slope of the tangent (rate of change) at that point.

Integrate
Differentiating gives a straight line (the tangent), and the inverse process (integrating) will go back to the curve.

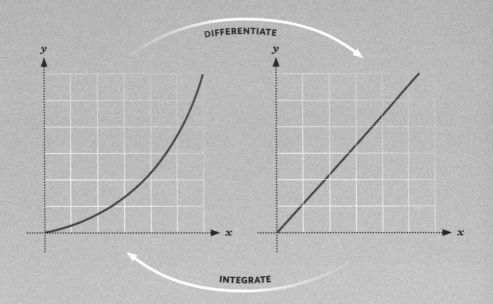

DIFFERENTIATE

INTEGRATE

INVERSE PROCESSES

Differentiation and integration are the processes of calculus. Differentiation determines the rate of change. Integration determines the possible original function if the rate of change is known. This means that differentiation and integration are inverse processes that undo one another. Integration usually results in a set of possible functions and will only give the original function if a point on the function is known. This is because the rate of change of a function is independent of its position relative to the x-axis. Differentiation always yields a single derivative function.

PRACTICAL CALCULUS

Calculus can be used to model situations involving quantities that change. Instead of building models and testing them, modern engineering practice often simulates them in a computer model. This allows greater control of the testing conditions and is more efficient (and safer) than physical models. Calculus also allows us to solve common optimization problems, such as working out the least amount of materials to make a can of drink with any given volume or optimizing the shape of a drinks can based on other criteria, such as how many cans fit onto a shelf.

Product design
These three cans have the same approximate volume, but use different amounts of material to make. The can that uses the least material is wider and harder to handle, whereas more of the slimline cans can be stacked on a shelf.

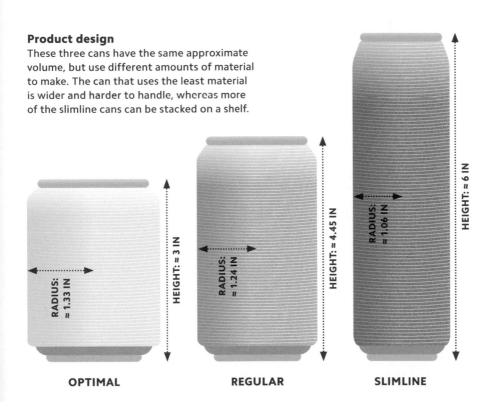

RADIUS: ≈ 1.33 IN
HEIGHT: ≈ 3 IN
OPTIMAL

RADIUS: ≈ 1.24 IN
HEIGHT: ≈ 4.45 IN
REGULAR

RADIUS: ≈ 1.06 IN
HEIGHT: ≈ 6 IN
SLIMLINE

INDEX

Page numbers in **bold** refer to main entries.

curves
  area under **152–53**
  bell 145
  parabolas **83**
cylinders 50, 107, 108

D

data
  analysis 130, **134–35**
  collection 129, 130, **131**, 132
  experimental probability 141
  handling cycle 130
  interpretation 130, **136**
  representation of 130, **132–33**
days 103
decagons 43
decibels 102
decimal point 12, 32
decimals **12**
  calculating **32–33**
  and proportion 90, 91
  rounding 34
degrees 40
denominators 11, 36–37, 67
density 121
derivative of a function **151**, 153, 154
Descartes, René 54
Devi, Shakuntala 20
diameter 105
dice 139, 140, 141
differentiation 149, 152, **154**
direct proportion **94–95**
direction 124–25
distance 102, 119, 121, 124–25
distance-time graphs 85
distribution graphs 145
division 27, **31**
  decimals 32–33
  laws of indices **68**
dodecagons 43
dodecahedrons 51

E

Earth, rotation and orbit 103
Einstein, Albert 60
electromagnetism 39
elevations **52**
elimination 73
encryption 19
engineering 22, 65, 155
enlargement **56–57**, 127
equations
  exponential 98
  linear **71**
  plotting **82–83**
  quadratic 54, **72**, 83, 84, 96
  simultaneous **73**
  solving with graphs **84**
equator 54
equilateral triangles 44
Eratosthenes 15
estimating **34–35**
  areas **152**
Euclid 39, 96
Euler, Leonhard 58
Euler's number 23
events
  Minkowski space 60–61
  probability 140–47
expanding **69**
experimental probability **139**, **141**, 146
exponentials 87, **98–99**
expressions, algebraic **66**
  expanding and factorizing **69**
  simplifying **67**
exterior angles 48, 49

F

factor trees 18
factorizing **69**
factors **14**, 15, 18–19, 37
Fibonacci 76, 77

Fibonacci sequence 76–77, 96
financial applications 15, 19, 65, 93
force 121
formulae **70**
fractal geometry 7, **62–63**
fractions **11**, 13, 23
  algebraic 67
  calculating **36–37**
  and proportion 90, 91
frequency 132, 133
functions 72, **80–81**
  derivatives **151**, 154
  integrals of **153**, 154

G

geographical coordinates 54
geometric sequences 75
geometry **38–63**
gigabytes 17
Global Positioning Systems (GPS) 118, 119
golden ratio 7, 76, **96–97**
gradient (graphs) 82, 150
graphs **78–85**
  data presentation **132–33**
  estimating areas **152**
  quadratic equations 72
  scatter 133, 137
greater than 74
greater than or equal to 74
group theory **25**
growth, exponential **98–99**

H

height 102
Heisenberg, Walter 45
hendecagons 43
heptagons 42
hexagons 42, 49
highest common factor 18, 89
horizontal shear 127

ACKNOWLEDGMENTS

DK would like to thank the following for their help with this book: Jessica Tapolcai and Mik Gates for the illustrations; Joy Evatt for proofreading; Helen Peters for the index. Senior Jacket Designer: Suhita Dharamjit; Senior DTP Designer: Harish Aggarwal; Jackets Editorial Coordinator: Priyanka Sharma.

Reference sources
90–91: Credit Suisse Global Wealth Report 2021 / credit-suisse.com/about-us/en/reports-research/global-wealth-report.html

All images © Dorling Kindersley
For further information see:
www.dkimages.com